AF799953

# Composites Science and Technology

**Series Editor**

Mohammad Jawaid, Department of Chemical and Petroleum Engineering, United Arab Emirates University, Al Ain, United Arab Emirates

This book series publishes cutting edge research monographs comprehensively covering topics in the field of composite science and technology. The books in this series are edited or authored by top researchers and professional across the globe. The series aims at publishing state-of-the-art research and development in areas including, but not limited to:

- Conventional Composites from natural and synthetic fibers
- Advanced Composites from natural and synthetic fibers
- Chemistry and biology of Composites and Biocomposites
- Fatigue damage modelling of Composites and Biocomposites
- Failure Analysis of Composites and Biocomposites
- Structural Health Monitoring of Composites and Biocomposites
- Durability of Composites and Biocomposites
- Biodegradability of Composites and Biocomposites
- Thermal properties of Composites and Biocomposites
- Flammability of Composites and Biocomposites
- Tribology of Composites and Biocomposites
- Applications of Composites and Biocomposites

**Review Process**

The proposal for each volume is reviewed by the main editor and/or the advisory board. The chapters in each volume are individually reviewed single blind by expert reviewers (at least two reviews per chapter) and the main editor.

**Ethics Statement** for this series can be found in the Springer standard guidelines here - https://www.springer.com/us/authors-editors/journal-author/journal-author-helpdesk/before-you-start/before-you-start/1330#c14214.

Showkat Ahmad Bhawani · Aamir Hussain Bhat ·
Mohamad Nasir Mohamad Ibrahim
Editors

# Biobased Materials and Their Composites for Oil Spill Treatment

Springer

*Editors*
Showkat Ahmad Bhawani
Faculty of Resource Science and Technology
Universiti Malaysia Sarawak
Kota Samarahan, Malaysia

Aamir Hussain Bhat
Department of Chemistry
College of Applied Sciences and Pharmacy
University of Technology and Applied Sciences
Muscat, Oman

Mohamad Nasir Mohamad Ibrahim
School of Chemical Sciences
Universiti Sains Malaysia
Minden, Malaysia

ISSN 2662-1819 ISSN 2662-1827 (electronic)
Composites Science and Technology
ISBN 978-981-96-6048-3 ISBN 978-981-96-6049-0 (eBook)
https://doi.org/10.1007/978-981-96-6049-0

© The Editor(s) (if applicable) and The Author(s), under exclusive license to Springer Nature Singapore Pte Ltd. 2025

This work is subject to copyright. All rights are solely and exclusively licensed by the Publisher, whether the whole or part of the material is concerned, specifically the rights of translation, reprinting, reuse of illustrations, recitation, broadcasting, reproduction on microfilms or in any other physical way, and transmission or information storage and retrieval, electronic adaptation, computer software, or by similar or dissimilar methodology now known or hereafter developed.
The use of general descriptive names, registered names, trademarks, service marks, etc. in this publication does not imply, even in the absence of a specific statement, that such names are exempt from the relevant protective laws and regulations and therefore free for general use.
The publisher, the authors and the editors are safe to assume that the advice and information in this book are believed to be true and accurate at the date of publication. Neither the publisher nor the authors or the editors give a warranty, expressed or implied, with respect to the material contained herein or for any errors or omissions that may have been made. The publisher remains neutral with regard to jurisdictional claims in published maps and institutional affiliations.

This Springer imprint is published by the registered company Springer Nature Singapore Pte Ltd.
The registered company address is: 152 Beach Road, #21-01/04 Gateway East, Singapore 189721, Singapore

If disposing of this product, please recycle the paper.

# Contents

# Introduction to Oil Spill Treatment

**Tarek Al-Arbi Omar Ganat**

**Abstract** The article gives a complete introduction to petroleum and explains why it is such an important energy source, but oil spills may result during transport. It focuses on the environmental harm caused by oil spills and examines a variety of factors that determine effects, such as the volume of the spill, type of oil involved, and geographical location. The paper explores existing oil spill recovery and treatment techniques such as mechanical, chemical, or biological ones. Their shortcomings are also pointed out therein. The debate also extends to the destiny of spilled oil. Natural processes involved in it include evaporation, dispersion and daily dissolution (depending on circumstances), remaining materials, degradation, etc. Technology can help man manage oil spills, but there is no technological solution to the problem of preventing them. Several case studies are given, looking at the long-term impact of oil spills on ecology. Moreover, the article discusses economic costs related to oil spills in terms of cleanup expenses, damage done by erosion of natural resources, and various losses suffered by impacted communities. Cost estimation factors such as oil quality. spill volume, location, weather conditions, and local response capacity are explored in detail herein. The study ends with a discussion of future perspectives, emphasizing such modern technologies as nanotechnology and bioremediation. The necessity of interdisciplinary research in environmental areas and the prospects for environmentally friendly materials are noted. The paper also points out the current limitations and difficulties in the prediction of best techniques for furthering biodegradation, as well as some doubts about introducing nanomaterials into marine environments. A comprehensive review finally points out that more research and determination of cost will be required to estimate the efficiency and sustainability of nanomaterials—as another cure for oil spill cleanup on a larger scale.

T. A.-A. O. Ganat (✉)
Petroleum and Chemical Engineering Department, Sultan Qaboos University, Muscat, Oman
e-mail: t.ganat@squ.edu.om

© The Author(s), under exclusive license to Springer Nature Singapore Pte Ltd. 2025

S. A. Bhawani et al. (eds.), *Biobased Materials and Their Composites for Oil Spill Treatment*, Composites Science and Technology,
https://doi.org/10.1007/978-981-96-6049-0_1

## 1 Background

The world's most significant energy source is petroleum. Numerous petrochemical processes, such as those that produce plastics and synthetic materials, require petroleum. Over the past few years, there has been a considerable improvement in oil output, and vast volumes of petroleum are constantly transported via trucks, trains, ships, and pipelines. The majority of this volume, along with its refined products, are shipped all over the world for different uses. Consequently, there is a chance that any of these modes of conveyance will unintentionally leak. There is always a chance of an oil spill, and they can seriously harm the ecosystem.

The consequences resulting from an oil spill (as depicted in Fig. 1) are contingent upon several factors, including the spill's volume, amount, type, and location. Over periods ranging from days to years, even a small spill can have a significant negative impact on individual species and entire populations, depending on the timing and location.

Spills typically occur when tankers or drilling platforms malfunction, when inland waterways contain boat traffic when wells blow out or pipelines break, or when fresh water is released from a well blowout or ruptured pipeline. Also, some huge oil spills have resulted from deliberate acts of warfare such as the Gulf War (Fig. 2). Where, during Operation Desert Storm (1991), hundreds of oil wells were burned in Kuwait, causing an environmental risk. Carbon from burning oil is a common component of soot particles found in oil well fire smoke. The smoke may also contain other particles, such dust and sand. As seen in Fig. 3, gases released by burning oil fires include carbon dioxide, carbon monoxide, acidic gases, sulphur oxide, hydrogen sulphide, nitrogen oxides, and volatile organic hydrocarbons.

**Fig. 1** Oil spill. *Source* https://news.mit.edu

**Fig. 2** Damaged oil well in Kuwait. *Source* www.usfca.edu

**Fig. 3** Oil field fires, Kuwait 1991, Earth from space. *Source* www.nasa.gov

Figure 4 uses data from the Coast Guard to show the number of occurrences and the amount of oil spills that occurred between 2002 and 2016. The spill data includes events from tankers, pipelines, and surface facilities. Since it is hard to compare the Deepwater Horizon oil leak to other accidents due to the spill's volume exceeding 100 million gallons, it is not included in the plot. Furthermore, the footage depicts the estimated 400,000 gallons of surface oil discharged by the Deepwater Horizon mobile offshore drilling rig [1].

With a few significant outliers in 2005 and 2006, the amount of oil flowing has remained generally constant over time, despite a drop in the number of breaches.

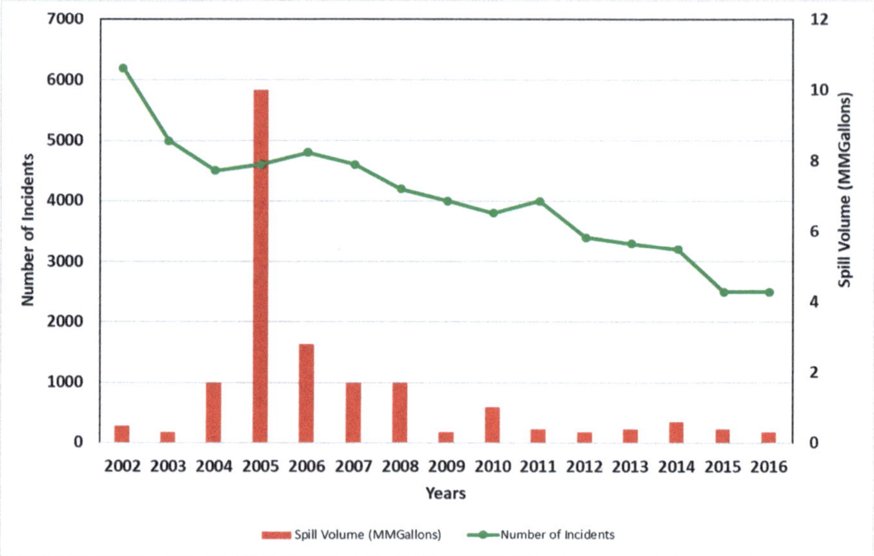


**Fig. 4** Shows the amount of oil spill accidents and spill volume between 2002 and 2016

Spills from 2005 surface facilities damaged by Hurricane Katrina in Louisiana totaled around 8 million gallons. Refined products spilled the following year account for nearly three-quarters of this figure (around US $436million) [2].

Oil spills continue to be a challenging problem to manage and control despite excessive advances in technology [3]. No cleaning method has proven to be 100% reliable and efficient, spills that are not cleaned up properly can contaminate layers around drinking water sources by seeping into the ground. They say cleaning up too often won't get rid of contaminants that contaminate the air, water, and soil after the spill. Post-spill cleanup efforts have consistently proven to be ineffective, expensive, and very difficult and impossible to fully rehabilitate affected areas. Thus, the most crucial approach to controlling oil pollution is to avoid oil leaks [4].

Case studies of particular spills and their consequences have been the main focus of earlier research. Ecosystems suffered for decades while oil spills were being cleaned up, according to Solomon and Janseen's (2010) analysis of the short- and long-term effects of the Deepwater Horizon oil spills [5]. Kujawinski et al. (2011) investigated the causes behind the deep-sea disaster (2010) and produced important findings and recommendations for the design and operation of upcoming deep wells. Most of these investigators chose oil spills, investigated various aspects of the disaster, and sought to obtain new data to clean up and prevent future oil spills. Therefore, there is a wealth of case studies in previous studies working on oil spill purification and containment strategies [6].

## 1.1 *Fate of Spilled Oil*

Many natural processes affect spillage after being released into the environment (Fig. 5). A lot of hydrocarbon fractions evaporate, dissolve in water, aggregate as residues, or break down in the presence of bacteria and sun UV radiation, depending on their chemical and physical characteristics:

### 1.1.1 Evaporation

Evaporation of vapors is essential to reduce the amount of residue that enters the environment. Normally, the evaporation process dissolves nearly 100% of gasoline leaked into the marine, 30–50% of crude oil, and 10% of bunker fuel. Light fractions usually evaporate, leaving heavy residues.

### 1.1.2 Dispersion

The term "dispersion" describes how oil slicks travel across open land or ocean surfaces. The spread can be very large. However, due to the high absorption capacity of the soil, it is much better controlled on land. The degree of dispersion on the surface of the water depends on the viscosity of the material and the ecological effects for instance wind speed, water flow, and the existence of surface ice.

### 1.1.3 Dissolution

Dissolution usually causes a waste of toxic water under the oil slick. Water solubility of light hydrocarbon fractions is higher than that of heavy fractions, and aromatic compounds are much more soluble than alkanes (Table 1).

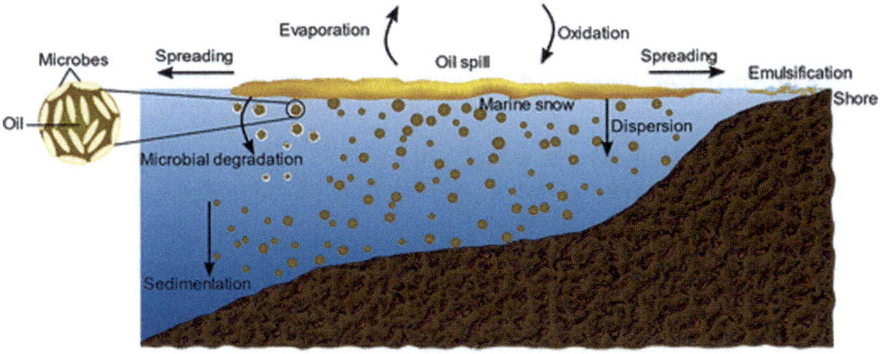


**Fig. 5** Fate of spilled petroleum on water [7]

**Table 1** Solubility of alkane and aromatic hydrocarbon in fresh water, g/m$^3$ (ppm)

| Hydrocarbon | Solubility |
|---|---|
| Gases | 24–62 |
| Liquids | 0.05–39 |
| Kerosene | $1–2 \times 10^{-4}$ |
| Lubricating oils | Less than $10^{-7}$ |
| Residual hydrocarbons | Less than $10^{-14}$ |
| Benzene | 1780 |
| Toluene | 515 |
| Naphthalene | 31 |

### 1.1.4 Remaining Materials

The residue remains after the light hydrocarbon fraction evaporates or dissolves. At sea, the remaining material forms a jelly-like water-in-oil emulsion called a "mousse." The oil that spills offshore is usually washed off the coast as a mousse and can then weather and deposit on rocks. Instead, the mousse can combine with shore sediment particles to make sticky tar-like patties that are eventually buried.

### 1.1.5 Degradation

The decomposition process means that bacteria, photooxidation, and solar ultraviolet radiation slow down the decomposition of substances. The energy source hydrocarbons are available to a wide variety of bacteria, fungi, and other microbes. But the rate at which biodegradation occurs depends significantly on ambient temperature, oxygen concentration, and availability of important nutrients. In general, light hydrocarbon fractions are generally easily broken down by inorganic and biological oxidation, whereas heavy fractions withstand this process and may remain rather stable in the environment [8].

## *1.2 Oil Spill Recovery and Treatment*

Preventing the pollution of surface and groundwater is the optimal approach in the event of an onshore oil spill. Onshore oil spill control and mitigation technologies include:

1. Soil berms, sorbent fences, or trenches can be used to enclose the soil and prevent oil from flowing into it.
2. The contaminated groundwater is surrounded or moved forward by slurry barriers constructed into a trench dug vertically.
3. Inject a viscous liquid barrier into the subsoil to separate contaminants.

The spread of contamination also depends on how quickly cleanup measures are initiated. There is greater accuracy in the movement and spill consequences forecast than the underwater environment, making it easier to plan an action or response approach. Slow weathering of soils and the harsh criteria generally set for soil remediation are other factors that help clean up spilled oil [9].

Approaches to oil spill response can be divided into mechanical, chemical, and biological [10]. Mechanical access includes a boom and a fixed flotation device that prevents the movement of oil spills. A hot boom allows for on-site burning of oil [11–13]. A skimmer is a stationary or mobile device used to collect emulsified oil suspended from the water surface [14]. A hydrophobic mesh that prevents water ingress but allows oil to pass through can be used on the skimmer for improved oil recovery. Chemical methods include spraying the oil spill with a dispersant [15] to break it down into droplets [16]. For small oil leaks, the sorbent can also be used. Adding microorganisms, nutrients, or oxygen to promote microbial development and the ensuing biodegradation of the spilled oil are examples of biological techniques [17]. On the other hand, physicochemical approaches have restrictions in terms of crude oil refining, thus biological approaches have been replaced in recent years [18]. However, the choice of adsorbent material depends on the nature of the leaked oil. The method of treatment determines how quickly oil solidifies in open water. Skimming and in situ combustion come next, with dispersants in air being the fastest [19]. The effectiveness of the purification approach is also affected by ecological conditions for instance wind, flow, and oil viscosity. The costs of various oil spill response approaches are on-site combustion, dispersants, mechanical recovery, and manual cleanup. Although cost-effective, in situ combustion produces large amounts of particulate matter and carbon dioxide in the atmosphere. The selection of an oil spill remediation technology becomes an optimization challenge by decreasing both the environmental impact and the process cost.

In general, the best solution for an oil spill will rely on several factors, and from an ecological and economic point of view, the best solution may be a combination of approaches. For oil to biodegrade or be collected from the environment, it must be left in the ecosystem in the event of an oil spill. Figure 6 shows the classification of various methods and materials.

## 2 The Current Oil Spill Treatment Techniques

Oil spill management includes assessing the amount of oil spilled into seawater, identifying the most effective cleaning techniques and evaluating the immediate and long-term effects of oil spills on the environment. There are several approaches to oil spill response, both onshore and offshore. Here are the most used methods:

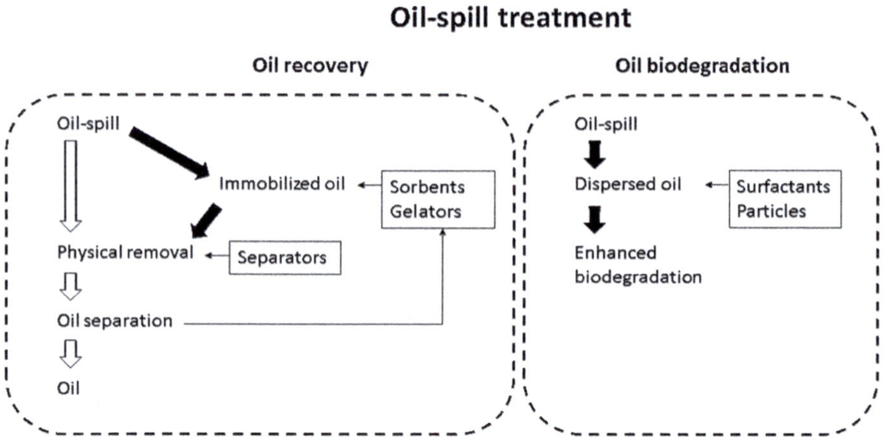


**Fig. 6** Treatment strategies for oil spills that take into account oil dispersion or oil recovery

## *2.1 Mechanical Methods*

The initial step in responding to an oil spill is to contain and gather spilled oil from the water's surface using mechanical techniques. These processes include booms, skimmers, and adsorbents (Fig. 7).

**Fig. 7** Sorbents oil spill response. *Source* www.oilspillresponse.gr

**Fig. 8** Two boats pull and band the oil spill using the inflatable boom. *Source* www.gloster.com

### 2.1.1 Booms

Mechanical methods are usually responsible for localizing the diffusion of oil. A boom, a floating barrier made from a mixture of materials, allows for effective and rapid oil removal at various oil viscosities but is relatively expensive. There are numerous kinds of booms, but all of them are physically oiled and all require a secondary approach to remove the oil. The protection provided by the boom enhances the effectiveness of other mechanical approaches [3]. Fireproof booms can be used to store oil to be burned on-site [5].

The boom is effective for cleaning when oil has accumulated in one place. The boom is used to protect the area from oil to prevent the oil from dispersing. There are three pieces to the boom, which floats on the water's surface. Spills are prevented from travelling down the boom by a "skirt" that is positioned beneath the water" surface and by the wire binding the arrow. The components of the boom are linked to the location of the oil spill (Fig. 8).

### 2.1.2 Skimmers

The skimmer is removed when oil is in the boom (Fig. 9). Skimmers collect oil from the water and either clean it or mentally separate the oil from the water and store it in a tank at the site of the spill. A large amount of oil can be extracted with the help of a skimmer.

Lipophilic, suction, and weir skimmers are the three different kinds of skimmers. Oil is successfully removed from the water's surface by the suction skimmer. A lipophilic skimmer uses straps, and chains of oleophilic, to remove spillage. Weir skimmer collects oil by gravity [6]. This method is best suited for marine spills.

**Fig. 9** Oil response weir skimmers. *Source* www.oilspillresponse.gr

Skimmer technology has not advanced significantly since the 1970s [7]. Skimmers are one of the fastest ways to clean up spilled oil [8]. Skimmers are not effective in large oil spills in rough seas or seawater. The skimmer process is labor intensive and can be harmful to health. Requiring labor can make skimmers expensive and clogging makes them less efficient [9].

### 2.1.3 Sorbents

Adsorbents are used to absorb oils in absorption or adsorption methods [15]. Oil is attracted to the surface of certain substances that are removed from the water. Sorbents are typically used to extract trace oils closer to shore or swamps [10]. Low-cost natural organic materials like moss, straw, feathers, cotton, hay, etc. are known as absorbents. Beaches close to the spill location are usually cleaned using vermiculite and hay. The primary issue with using sorbents is that they become denser when coated in oil, which causes them to sink to the bottom of the ocean and damage the marine ecosystem. Typically, absorbents are employed to remove fat from an afflicted animal's body.

Synthetic materials are another source of adsorbents [11]. It is imperative to remove the oil-absorbing sorbent from the spill site as soon as it has been used. Otherwise, the adsorbent will sink to the sea or ocean floor [10]. These "spongy" materials with modified surface chemistry can provide fast and efficient oil removal from spill sites [12–14] (Table 2).

**Table 2** Materials that sorbent spilled oil

| Natural organic | Natural inorganic | Synthetic |
|---|---|---|
| Hay | Volcanic ash | Polypropylene |
| Feathers | Perlite | |
| Peat moss | Clay | Polyurethane |
| Straw | Glass wool | Polyethylene |
| Sawdust | Sand | Rubber |
| | Vermiculite | |

*Source* www.epa.gov

## 2.2 Thermal Methods

Burning oil spills typically require specific equipment, for example, igniters and oil can be contained with fire-resistant booms. Lighters fixed to the surface of boats or helicopters burn the oil enclosed within the boom. On-site burning of oil is one of the most effective and fastest ways to respond to oil spills, but there are certain periods during which on-site burning is effective because of the oil's natural emulsification [16]. For ignition, the spill thickness should be about 3 mm and the water content should be about 25% [17]. Removal can be as high as 95% when used in purified water until the oil film is emulsified in water. Negative consequences of burning oil on site are air pollution, residues can fall to the bottom of the sea, or create a secondary fire hazard [3, 18, 19].

## 2.3 Chemical Methods

There are many chemicals used to treat oil spills and reduce their impact. Oil spills are usually broken up into tiny droplets by emulsifiers and dispersants [3]. Instead, an oil skimmer is an oil recovery agent that collects the oil into gelled water that can be burned on site. Oil herders are silicone-based, non-biodegradable, and waterproof [17]. In general, these methods might be applied in all weather conditions where traditional mechanical approaches are ineffective [15]. Surfactants are also used to dissolve oil in water. Surfactants can increase the bioavailability of oils for microbial degradation by reducing the interfacial tension of oil and water and stabilizing emulsion droplets [20]. The effectiveness of Corexit has been studied in detail and studies have shown that it is an effective dispersant, but its presence may reduce bioremediation. Several of these findings have been disproved as a consequence of inconsistent laboratory research [21].

In recent oil disasters, toxic emulsifiers and surfactants were utilised. The long-term environmental effects of surfactants have been the subject of numerous studies, however the findings are currently contradictory.

### 2.4 Natural Methods

Bioremediation is usually a natural way to break down spilled oil by endemic microorganisms. Many species can break down oil, such as bacteria, fungi, and algae. Bioremediation is very effective and has become an important method in some cleanups. The efficiency of microbial degradation depends primarily on the bioavailability of oils and nutrients. The bioremediation-enhanced approach requires more growth-restricting nutrients, which can result in longer microbial delays and ultimately longer wash times. Bioremediation can be stopped mechanically and chemically. Because of their toxicity, chemical methods can stop microorganisms from degrading oils, whereas mechanical processes entail the localization of oils, which can impede their growth owing to a reduction in sunshine and oxygen [26].

### 2.5 Nonoparticles for Enhancing Bioremediation

Nanoparticles have been extensively studied as a potential option for treating oil spills throughout the last ten years. A number of these cutting-edge technologies have the potential to support native microorganisms' bioremediation efforts, and nanoparticles are an efficient way to remediate soil [27]. The development of these natural methods throughout training is essential. Currently, most research on nanobiological purification is still on the laboratory scale. Therefore, it is necessary to perform pilot tests in real field conditions. Currently, evaluations of oil spill cleanup tend to concentrate on a single procedure, like bioremediation or the application of nanomaterials [29, 30]. This assessment is the first to offer a significant examination of the application of nanomaterials to enhance bioremediation.

## 3 The Future Oil Spill Treatment Methods

There is great demand for an innovative approach to extract oil from water mixtures and recover it. Current technology is still in use, but much research has been done to optimize conventional technology and develop new ways to speed up and improve oil spill treatment processes. As a result, numerous research projects have been undertaken to enhance the mixture's separation. A topcoat of fluorine-based surfactants was coated with a nanoparticle/binder coating by Bhushan and Martin [20] to give water affinity and oil repellency; roughness increases these qualities to oil over 150°. It provides a contact angle and a water contact angle of less than 5°. When applying this coating to a stainless-steel mesh, you can pour a mixture of water and oil over the mesh, allowing the water to flow through it easily while the oil remains on top. By tilting the screen, the liquid can be separated and recovered. Note that the nanoparticle/binder complex coating method works efficiently on a set of substrates with favorable performance [20]. It requires some processing steps, which is desirable [21].

The efficacy of *O*-carboxymethyl chitosan and the sodium form of amphiphilic chitosan as adsorbents for oil removal and recovery from surface water was examined by Doshi et al. [21, 22]. Water-soluble was synthesized by carboxymethylation of chitosan in the hydroalcoholic medium at 50 °C using chloroacetic acid. By adding hydrophobic groups to the amine groups of the hydrophilic sodium carboxymethyl chitosan salt, the sodium oleoyl carboxymethyl chitosan salt is created. The research examined the efficacy of NOCS while accounting for the concentration, acidity, and pH of oil2T, marine diesel, and marine diesel. Out of the three oils that were investigated, the results indicate that marine diesel oil has the most effective destabilization because of its exceptional adsorption capacity to both the salinity and alkalinity of seawater. Droplets smaller than 30 μm for six weeks. Recovering oil from deionized water takes 75–85%, while recovering oil from seawater takes 19–49%. Surface wetting for spill separation was achieved by Gupta et al. [23] using a modified superhydrophobic/superhydrophilic Janus membrane. They impregnated a nano-sized polytetrafluoroethylene spread on a cotton bed using the Meyer rod coating method. The membrane shows excellent separation efficiency up to 98%. The use of filter separation techniques in oil spill treatment on land has a wide perspective. Turco et al. [24] presented an innovative method to produce porous polydimethylsiloxane from a water-silicone reverse process capable of selectively aggregating oil from water in a short time. The results demonstrate that the absorbance of the oil mass is between 3.9 and 32.3 g/g. This method works in emergencies, allowing on-site production of porous absorbents. Porous materials show higher volumetric capacity to absorb oil from water.

Furlan et al. [25] produced a high-performance polydimethylsiloxane magnetic nano-synthetic sponge material for spill handling. The findings show the adsorption capacity for cleaning agents (440%), emulsifying oil drops (580%), and oils with varying densities and viscosities (415–760%) as determined by weight growth. According to the results, the magnetic sponge can also selectively adsorb oils in oil–water emulsions and has a high adsorption capacity for oils and detergents. It is also resistant to harsh chemicals. Regular usage of sponges won't cause any wear and tear.

Spills can be treated with suspended microbial cotton using activated sludge, an aerobic biological treatment method. A batch sequencing reactor's full cycle consists of five stages: no-load, water entry, reaction, sedimentation, and water exit. Because of its great efficiency and cheap running costs, sequential batch reactors are used extensively in wastewater treatment. Recently, Uma et al. [28] investigated the treatment of synthetic oily seabed water and subsequent biopolymer production using a sequencing batch reactor with newly isolated soil bacteria from the site of contaminated hydrocarbons near Port of Karaikal, India. Experimental results are shown to reduce substrate absorption at 5000 mg/L dissolved chemical oxygen demand indicating maximum organic matter removal (81%).

# 4 Economic Costs of Oil Spills

1. Oil spills are a very common case. Cost estimation and classification is difficult and there is a lake of data. Consequently, the computation's accuracy is not fully accurate. But for regular enterprises, port managers, and environmental managers, these costs need to be calculated. This aids in categorizing the necessary funding routes in the event of an oil spill at or close to a port.
2. The possible economic costs of oil spills fall into three categories: cleanup costs, damage to natural resources, and various economic losses to affected communities and individuals. A 2012 study of the economic impact of offshore vessel spills found that medium-sized oil spills were estimated at approximately $189 million and cleanup costs at $2.4 billion. A major oil spill cost about $308 million and cleanup would require $9.4 billion [30].

## *4.1 Factors Affecting the Estimation of Oil Spills*

The calculation of environmental expenses and damage resulting from oil spills is influenced by the following elements:

### 4.1.1 Oil Quality

Complex blends with a range of characteristics make up oils. Physical and chemical alterations occur in oil as it rises to the ocean's surface. These methods are dissolution, diffusion, evaporation, and precipitation. The increase in the structure and viscosity of the oil-in-water emulsion contributes to the stability of the oil in the environment. Marine environments absorb oil through biodegradation processes for a long time. In general, diesel does not stay at sea level for long. Therefore, the cost of refining these diesel fuels is low. However, heavy oil is more expensive because it contains non-volatile components.

In general, light oil is a fire hazard and has a higher environmental toxicity than heavy oil. Heavy oil tends to accumulate in bottom sediments.

### 4.1.2 Oil Volume

The amount of oil has a big impact on the cleaning budget. The amount that spills expands and moves more when it is larger, and the expense of purification increases proportionally.

#### 4.1.3 Site of the Oil Spill

The location of an oil spill has a significant impact on cleanup costs. Location is important to rescue teams and the extent of damage to the environment. Purification is different for land and sea events. A simple approach is usually used on land. However, the marine sector requires very sophisticated equipment, so the cost of purification is higher. On the other hand, it is easier to clean up oil spills at sea than at shore.

#### 4.1.4 Weather and Sea Conditions

Marine and climatic factors play a significant role in lessening the effects of oil spills. The temperature influences how quickly the head is cleaned. In some cases, spills took nearly a year to complete. The main cause of this long period is the weather conditions.

#### 4.1.5 The Local Response Capacity

Rescue operations depend on the availability of equipment such as chemicals, buoys, airplanes, lifeboats, etc. These gadgets are typically not always readily available, since it depends on the local government and attitude in the area. For example, if an incident occurs near a tourist destination, pressure to clean up an oil spill can be high due to pressure from media attention. On the other hand, if oil spills happen in areas like salt marshes or mangroves, they will be cleaned up naturally. The decision was made to leave the oil that had leaked on the rocky shore after numerous previous oil accidents.

### *4.2 Other Economic Costs*

Damage to natural resources or other costs beyond operating expenses might result from oil spills. Businesses and people that depend on the surrounding environment are the major types of economic activity that can be disrupted by oil spills close to the spill location. Local tourism and fishing, for instance, may be impacted. Even though there is a genuine risk of oil spills to human health, there are instances when well-publicized oil disasters cause regional companies near the spill site to fail.

In addition, in the case of an oil leak, local services can be interrupted. Furthermore, port operations might be suspended, which would change the movement of commercial commodities. Power plants could need to be shut down.

## 5 Future Perspectives and Conclusions

Current approaches to oil spill cleanup are discussed in this introduction, along with advancements in nanoscience and bioremediation. Nanotechnology-based biological treatment is a novel field of study that requires more investigation. The transdisciplinary suit presents novel features and inventive options, like the greater dispersal of microorganisms on oil droplets rather than droplets and the utilisation of new environmentally safe and sustainable materials. harmful chemicals and remove petroleum with magnets instead of skimming agents and absorbents. In addition, there are some limitations and challenges such as:

1. It is difficult to predict the optimal technique for enhanced biodegradation because of the wide range of temperatures, salinities, pH and other ecological conditions
2. Biological effects of introducing nanomaterials into the marine location.
3. Process of separating nanomaterials from seawater.

Subsequent research ought to concentrate on eco-friendly substances like clay and organic biofilm-forming agents. Some materials can be very effective as Pickering emulsifiers in enhancing microbial propagation and they have low toxicity. This new method will provide better biodegradability of spills, increase oil recovery, and permanently reduce economic and ecological impacts at spill sites. Research on employing nanotechnology to clean up oil spills has significantly increased recently. The best method for removing oil through a natural process is biological therapy. Furthermore, a small number of studies have coupled nanomaterials and bioremediation to enhance the cleanup of oil spills, and the outcomes are encouraging. Using Pickering emulsifiers to increase the pace of breakdown and promote microbial dispersion on oil droplets is probably the most likely breakthrough in oil spill response. Mesocosm studies are used in additional research to establish the efficacy of nanoparticles in bioremediation. It is important to show this efficacy in settings that change over time, such as pH, temperature, salinity, and agitation. Moreover, it is not feasible to use nanoparticles in the event of the next significant oil spill. By combining this field with bioremediation, a connection between laboratory- and industrial-scale consumption might be made. In the end, these investigations validate nanoparticles as a low-cost substitute; nevertheless, no official data exists to evaluate the expenses associated with traditional methods of utilizing nanomaterials for large-scale applications. To demonstrate that nanoparticles are unquestionably a viable and economical substitute for traditional oil spill cleanup techniques, a cost analysis must be carried out.

## References

1. Whitehead B (1994) The off-site rule. CERCLA information brief. USDOE assistant secretary for environment, safety, and health, Washington, DC (United States). RCRA/CERCLA Div

2. Wenning RJ, Robinson H, Bock M, Rempel-Hester MA, Gardiner W (2018) Current practices and knowledge supporting oil spill risk assessment in the Arctic. Mar Environ Res 1(141):289–304
3. Absollhi M (ed) (2014) Encyclopedia of toxicology. Elsevier
4. Onwurah IN, Ogugua VN, Onyike NB, Ochonogor AE, Otitoju OF (2007) Crude oil spills in the environment, effects and some innovative clean-up biotechnologies
5. Solomon GM, Janssen S (2010) Health effects of the Gulf oil spill. JAMA 304(10):1118–1119
6. Kujawinski EB, Kido Soule MC, Valentine DL, Boysen AK, Longnecker K, Redmond MC (2011) Fate of dispersants associated with the Deepwater horizon oil spill. Environ Sci Technol 45(4):1298–1306
7. Clark RC, MacLeod WD (1977) Inputs, transport mechanisms, and observed concentrations of petroleum in the marine environment. Effects Petrol Arctic Subarctic Mar Environ Organ 1:91–223
8. Socolofsky SA, Gros J, North E, Boufadel MC, Parkerton TF, Adams EE (2019) The treatment of biodegradation in models of sub-surface oil spills: a review and sensitivity study. Mar Pollut Bull 1(143):204–219
9. Acosta-González A, Martirani-von Abercron SM, Rosselló-Móra R, Wittich RM, Marqués S (2015) The effect of oil spills on the bacterial diversity and catabolic function in coastal sediments: a case study on the prestige oil spill. Environ Sci Pollut Res 22:15200–15214
10. Board OS, National Academies of Sciences, Engineering, and Medicine (2020) The use of dispersants in marine oil spill response. National Academies Press
11. Evans DD, Mulholland GW, Baum HR, Walton WD, McGrattan KB (2001) In situ burning of oil spills. J Res Nat Inst Stand Technol 106(1):231
12. Fingas M (2011) An overview of in-situ burning. Oil Spill Sci Technol 1:737–903
13. Wolok E, Barafi J, Joshi N, Girimonte R, Chakraborty S (2020) Study of bio-materials for removal of the oil spill. Arab J Geosci 13:1–1
14. Michel J, Shigenaka G, Hoff R (1992) Oil spill response and cleanup techniques. An introduction to coastal habitats and biological resources for oil spill response. NOAA, Seattle, 401:1–3
15. Graham LJ, Hale C, Maung-Douglass E, Sempier S, Swann L, Wilson M (2016) Chemical dispersants and their role in oil spill response. Oil Spill Sci Sea Grant Programs Gulf Mex 1
16. Jensen JR, Halls JN, Michel J (1998) A systems approach to Environmental Sensitivity Index (ESI) mapping for oil spill contingency planning and response. Photogramm Eng Remote Sens 1(64):1003–1014
17. Azubuike CC, Chikere CB, Okpokwasili GC (2016) Bioremediation techniques–classification based on site of application: principles, advantages, limitations and prospects. World J Microbiol Biotechnol 32:1–8
18. Das N, Chandran P (2011) Microbial degradation of petroleum hydrocarbon contaminants: an overview. Biotechnol Res Int
19. Zhang T, Li Z, Lü Y, Liu Y, Yang D, Li Q, Qiu F (2019) Recent progress and future prospects of oil-absorbing materials. Chin J Chem Eng 27(6):1282–1295
20. Bhushan B, Martin S (2018) Substrate-independent superliquiphobic coatings for water, oil, and surfactant repellency: an overview. J Colloid Interface Sci 15(526):90–105
21. Doshi B, Repo E, Heiskanen JP, Sirviö JA, Sillanpää M (2017) Effectiveness of N, O-carboxymethyl chitosan on destabilization of marine diesel, diesel and marine-2T oil for oil spill treatment. Carbohyd Polym 1(167):326–336
22. Doshi B, Repo E, Heiskanen JP, Sirviö JA, Sillanpää M (2018) Sodium salt of oleoyl carboxymethyl chitosan: a sustainable adsorbent in the oil spill treatment. J Clean Prod 1(170):339–350
23. Gupta P, Kandasubramanian B (2017) Directional fluid gating by janus membranes with heterogeneous wetting properties for selective oil–water separation. ACS Appl Mater Interfaces 9(22):19102–19113
24. Turco A, Primiceri E, Frigione M, Maruccio G, Malitesta C (2017) An innovative, fast and facile soft-template approach for the fabrication of porous PDMS for oil–water separation. J Mater Chem A 5(45):23785–23793

25. Furlan PY, Ackerman BM, Melcer ME, Perez SE (2017) Reusable magnetic nanocomposite sponges for removing oil from water discharges. J Ship Prod Design 33(03):227–236
26. Zhang W, Wei Y, Jin Y (2018) Full-scale processing by anaerobic baffle reactor, sequencing batch reactor, and sand filter for treating high-salinity wastewater from offshore oil rigs. Processes 6(12):256
27. Choi YI, Ji HJ, Shin DY, Mansoor S, Kwan MJ, Lee SC, Jung JH, Jung BG, Sung NC, Wang JP (2019) An evaluation of the water quality characteristics of shipboard sewage disposal and usability based on water quality enhancement. Appl Sci 9(3):418
28. Uma V, Gandhimathi R (2019) Organic removal and synthesis of biopolymer from synthetic oily bilge water using the novel mixed bacterial consortium. Biores Technol 1(273):169–176
29. Martin S, Brown PS, Bhushan B (2017) Fabrication techniques for bioinspired, mechanically-durable, superliquiphobic surfaces for water, oil, and surfactant repellency. Adv Coll Interface Sci 1(241):1–23
30. Hotte N, Sumaila UR (2012) Potential economic impact of a tanker spill on ocean-based industries in British Columbia

# An Overview of Mechanized Solutions for Oil Spill Treatment

**Nadeem Akbar Najar, Imtiyaz Akbar Najar, Showkat Ahmad Bhawani, Farooq Ahmad Najar, and Farooz Ahmad Najar**

**Abstract** Oil spills pose significant environmental hazards and require efficient and timely response measures for containment and cleanup. Mechanized solutions have proven to be instrumental in combating oil spills, offering effective containment, recovery, and treatment methods. This chapter provides an overview of various mechanized solutions used in the treatment of oil spills. We explore key technologies, equipment, and strategies employed in the response to oil spill incidents, emphasizing their effectiveness, limitations, and future prospects.

**Keywords** Oil spill · Mechanized solution · Treatment · Environmental hazards

N. A. Najar
Centre for Policy Studies, Indian Institute of Technology Bombay, Powai Mumbai, Maharashtra, India

Mechanical Engineering Department, Government of Jammu and Kashmir, Srinagar, India

I. A. Najar (✉)
Faculty of Engineering, Universiti Malaysia Sarawak, Kota Samarahan, Sarawak, Malaysia
e-mail: imtiyaznajar999@gmail.com

S. A. Bhawani
Faculty of Resource Science and Technology, Universiti Malaysia Sarawak, Kota Samarahan, Sarawak, Malaysia

F. A. Najar
Department of Mechanical Engineering, Institute of Technology, University of Kashmir, Srinagar, India

F. A. Najar
Department of Mechanical Engineering, National Institute of Technology Srinagar, Srinagar, Kashmir, India

© The Author(s), under exclusive license to Springer Nature Singapore Pte Ltd. 2025

S. A. Bhawani et al. (eds.), *Biobased Materials and Their Composites for Oil Spill Treatment*, Composites Science and Technology,
https://doi.org/10.1007/978-981-96-6049-0_2

# 1 Introduction

Oil spills pose a significant threat to the environment, ecosystems, and human health, necessitating effective and timely remediation strategies [1, 2]. Over the years, numerous techniques and technologies have been developed to tackle the challenging task of oil spill treatment [3, 4]. Among these, mechanized solutions have emerged as a vital component in the arsenal of tools utilized to combat and mitigate the devastating impact of oil spills. In this chapter, we will provide a comprehensive overview of the mechanized solutions employed for oil spill treatment. We will delve into the various technologies, equipment, and methodologies that are utilized to effectively contain, recover, and clean up oil spills. By exploring the advancements in mechanized solutions, we aim to shed light on the evolution of oil spill response and emphasize the importance of innovative approaches to tackle these environmental disasters [5, 6]. The chapter will begin by highlighting the need for mechanized solutions in oil spill treatment, elucidating the limitations of manual intervention and the advantages offered by mechanized approaches. We will discuss the critical factors that influence the selection of appropriate mechanized solutions, such as the type and volume of the spilled oil, environmental conditions, and logistical considerations.

Subsequently, this chapter explored the various categories of mechanized solutions employed in oil spill treatment. This will encompass containment and recovery systems, skimmers, booms, sorbents, and dispersants, among others. Each category will be examined in detail, outlining the principles of operation, technological advancements, and their effectiveness in different scenarios.

Furthermore, this chapter examined case studies and real-world applications of mechanized solutions for oil spill treatment. These examples will provide insights into the successful implementation of specific technologies and highlight the challenges encountered during oil spill response operations. Additionally, we will discuss the associated environmental impacts and considerations when utilizing mechanized solutions, ensuring a holistic perspective on the topic.

Finally, this chapter assessed the current state of research and development in mechanized solutions, as well as emerging trends and future prospects. We will explore cutting-edge technologies, such as autonomous systems, drones, robotics, and artificial intelligence, which are revolutionizing oil spill treatment. By embracing these advancements, we can enhance our ability to respond effectively to oil spills and minimize their ecological consequences. By examining the technologies, methodologies, and advancements in this field, we hope to inspire further innovation, collaboration, and implementation of mechanized solutions to protect our oceans, coastal areas, and fragile ecosystems from the devastating impact of oil spills.

# 2 Various Techniques

Containment and recovery techniques (Fig. 1).

**Fig. 1** Oil booms used to contain oil spill (figure adapted from [7])

## 2.1 *Booms and Barriers*

Booms and barriers are commonly used for the containment of oil spills, preventing their spread and confining them to a localized area. These structures are typically made of durable materials, such as PVC, rubber, or fabric, and are designed to withstand harsh marine conditions [8, 9]. Booms come in various types, including inflatable, sorbent, and fence booms, each suited for different spill scenarios [7–10].

I. **Inflatable Booms**: These booms consist of air-filled chambers that provide buoyancy and create a physical barrier between the spilled oil and surrounding water. Inflatable booms are effective for containing spills in open water and can be deployed quickly. However, they are susceptible to damage from rough weather conditions and may require regular monitoring and maintenance [9].
II. **Sorbent Booms**: Sorbent booms are designed to absorb oil, using materials such as polypropylene or natural fibers. As oil contacts the boom, it is absorbed and retained within the sorbent material, allowing for recovery. Sorbent booms are particularly useful in calm water areas and can be utilized to remove oil sheens from the surface [11].
III. **Fence Booms**: Fence booms are vertical barriers that act as a physical obstacle to contain oil spills. They are often used in areas with strong currents or where the terrain restricts the deployment of traditional booms. Fence booms are constructed with sturdy materials and can be anchored to the seabed or secured to floating platforms [7].

## 2.2 Skimmers and Sorbents

Skimmers and sorbents are mechanical devices used to recover oil from the water surface [11, 12]. They are designed to separate the oil from the water, allowing for efficient oil recovery. Skimmers are specialized devices that skim oil from the water surface. They utilize different methods, such as suction, brushes, or rotating drums, to collect the oil. Skimmers can vary in size and capacity, ranging from small portable units to large vessels equipped with high-capacity skimming systems [2, 11, 12]. Skimmers are particularly effective for recovering thick layers of floating oil, but their efficiency may vary depending on the oil type and environmental conditions. Sorbents, such as pads, booms, or granular materials, are used to absorb oil from the water surface. These materials have a high affinity for oil and can effectively soak it up, allowing for subsequent recovery. Sorbents are commonly employed in combination with skimmers or manually deployed to absorb and remove oil from sensitive areas, shorelines, or wildlife habitats. However, they have a limited capacity for oil absorption and need to be replaced or disposed of properly [13, 14].

Effective containment and recovery techniques are critical during oil spill incidents. Booms and barriers act as physical barriers to prevent the spread of oil, while skimmers and sorbents aid in the removal and recovery of oil from the water surface [15]. The selection of appropriate techniques depends on factors such as spill size, location, weather conditions, and the type of oil involved [16, 17]. Integration of these techniques with other response measures, such as dispersants and shoreline cleanup [18], is often necessary to achieve comprehensive and effective oil spill treatment [11, 19]. Skimmers and sorbents are key tools used in the containment and recovery of oil spills. They play a crucial role in removing oil from the water surface, reducing the environmental impact and facilitating the cleanup process.

## 2.3 Skimmers

Skimmers are mechanical devices specifically designed to remove oil from the water surface. They utilize various mechanisms to separate oil from water, allowing for efficient recovery [2, 9, 11]. Here are some common types of skimmers:

I. **Suction Skimmers**: These skimmers use suction to remove oil from the water surface. They typically consist of a vacuum pump or an eductor system that creates a suction force. As the skimmer moves across the oil-contaminated area, the suction inlet draws in the oil, which is then transferred to a collection container for further processing. Suction skimmers are effective in collecting both light and heavy oils.

II. **Brush Skimmers**: Brush skimmers employ rotating brushes or discs to collect oil from the water surface. The brushes or discs are designed to have oil-adherent surfaces, which attract and retain the oil as they rotate through the water. The collected oil is then scraped off the brushes or discs and directed into storage tanks for recovery. Brush skimmers are particularly effective in removing thick layers of oil and can handle high viscosity oils.

III. **Drum Skimmers**: Drum skimmers consist of a rotating drum or a set of drums partially submerged in the water. The drum surface is made of a material that can efficiently pick up and retain oil as it rotates through the water surface. The oil is then scraped off the drum surface and collected for further processing. Drum skimmers are suitable for both light and heavy oils and are often used in open water or offshore spill scenarios.

## *2.4 Sorbents*

Sorbents are materials that have a high affinity for oil and are used to absorb or adsorb oil from the water surface. They can be in the form of pads, booms, pillows, or granular materials. Sorbents work by allowing oil to adhere to their surfaces, effectively removing it from the water [11, 14, 20, 21]. Here are some common types of sorbents:

I. **Absorbent Pads and Sheets**: These are flat, absorbent materials designed to soak up oil from the water surface. They are made of materials such as polypropylene or natural fibers that can effectively absorb and retain the oil. Absorbent pads are often used for small-scale spills or in areas where manual cleanup is required.

II. **Sorbent Booms**: Sorbent booms are cylindrical or elongated structures filled with sorbent material. They are used to contain and absorb oil in water bodies. Sorbent booms can be deployed around sensitive areas, shorelines, or wildlife habitats to prevent oil from spreading and to aid in the recovery process.

III. **Granular Sorbents**: Granular sorbents are small particles or granules made of materials such as clay, cellulose, or silica. They are spread over the oil-contaminated surface, where they absorb the oil, forming a solid or semi-solid mass. Granular sorbents are effective in treating small spills or when oil is trapped in hard-to-reach areas.

It is important to note that while skimmers and sorbents are effective in removing oil from the water surface, their efficiency can be influenced by several factors. These include the type and viscosity of the oil, sea and weather conditions, and the operational characteristics of the skimmer or sorbent material [4, 9, 11, 16]. Proper selection, deployment, and maintenance of skimmers and sorbents are essential for maximizing their effectiveness during oil spill response. Furthermore, the collected oil from skimmers and sorbents needs to be properly handled and disposed of in accordance with environmental regulations. Skimmers may require onboard storage tanks or transfer systems for efficient oil recovery, while sorbents may need proper disposal methods to avoid secondary pollution.

Skimmers and sorbents are valuable tools in the initial containment and recovery stages of oil spill response. They contribute to minimizing the environmental impact of oil spills and play a significant role in the overall cleanup process. However, it is important to employ a comprehensive approach that integrates skimmers, sorbents, and other oil spill response techniques to ensure effective and efficient treatment of oil spills. Mechanical separation and solidification techniques are effective approaches used in the treatment of oil spills. These methods focus on physically separating the oil from water or converting it into a solid or semi-solid form. Let us explore two common techniques: centrifuges and separators, as well as solidification and encapsulation.

## *2.5 Centrifuges and Separators*

Centrifuges and separators are mechanical devices that utilize the principle of centrifugal force to separate oil from water. These techniques are particularly effective when dealing with emulsions or oil–water mixtures that are challenging to separate using conventional methods.

I. **Centrifuges**: Centrifuges employ high-speed rotation to separate oil and water based on their density differences. The centrifugal force generated causes the denser oil phase to move towards the outer edge of the centrifuge, while the lighter water phase remains closer to the center. This separation allows for the efficient extraction of oil. Centrifuges are commonly used for treating oily wastewater and can be adapted for oil spill response, especially in cases where oil and water are strongly emulsified [22].
II. **Separators**: Separators, also known as hydrocyclones or hydro separators, use a combination of centrifugal force and gravitational settling to separate oil and water. The mixture enters the separator, and the high rotational speed creates a cyclonic motion that forces the oil to move towards the center and the water to move towards the periphery. This separation process allows for the collection and removal of oil from the water phase. Separators are often used in offshore oil production facilities and can be adapted for oil spill treatment [11].

Centrifuges and separators are effective in treating oil–water mixtures by providing a mechanical means of separation. However, they may require additional treatment steps, such as further filtration or oil–water separation processes, to achieve optimal results.

## *2.6 Solidification and Encapsulation*

Solidification and encapsulation techniques involve transforming oil into a solid or semi-solid form, preventing its further spread and facilitating its removal [23]. These methods are particularly useful for treating heavy oils or oil residues that are difficult to recover using conventional techniques.

I. **Solidification**: Solidification involves adding solidifying agents, such as clay, sawdust, or polymers, to the oil to convert it into a semi-solid mass. The solidified oil can then be manually collected or mechanically removed. Solidification is effective in immobilizing oil and preventing its migration to sensitive areas or shorelines. However, it may not be suitable for large-scale spills due to the significant quantities of solidifying agents required.

II. **Encapsulation**: Encapsulation techniques involve coating oil droplets with materials, such as polymers or surfactants, to form small capsules. These capsules prevent the oil from dispersing and facilitate its removal. Encapsulation is often used in conjunction with other treatment methods, such as skimming or sorbents, to enhance oil recovery. It can be particularly effective for treating oil in confined or hard-to-reach areas [24].

Solidification and encapsulation techniques offer advantages in terms of immobilizing oil and reducing its environmental impact [8]. However, it is important to consider the long-term stability of the solidified or encapsulated oil, as well as the potential for secondary pollution during the cleanup process. These methods provide effective options for treating oil spills and can be employed as part of a comprehensive response strategy to minimize the environmental impact of oil contamination. Solidification and encapsulation techniques are utilized in the treatment of oil spills to transform the spilled oil into a solid or semi-solid state, thereby facilitating containment and removal. These techniques are particularly effective in dealing with heavy oils, viscous residues, or situations where conventional recovery methods are challenging. Let us explore solidification and encapsulation in more detail.

## *2.7 Solidification*

Solidification involves the addition of solidifying agents to the spilled oil, causing it to undergo a physical or chemical change and form a solid or semi-solid mass [8]. The solidified oil can then be easily collected and removed. Here are key aspects of solidification:

I. **Solidifying Agents**: Various materials can be used as solidifying agents, including clay, sawdust, sand, vermiculite, or synthetic polymers. These agents have a high affinity for oil and can bind to the hydrocarbons, transforming them into a more manageable form [25].

II. **Application Process**: Solidifying agents are typically spread or applied over the oil spill area manually or by using specialized equipment. They are mixed or blended with the spilled oil, either on the water surface or on affected shorelines, depending on the spill location and characteristics [26].

III. **Benefits and Considerations**: Solidification prevents the spread of oil by immobilizing it, minimizing the risk of further contamination. It facilitates the collection and removal of the oil, reduces its toxicity, and helps prevent its interaction with sensitive habitats. However, solidification may require significant amounts of solidifying agents, and the disposal of the solidified oil should be done in compliance with environmental regulations [27].

## 2.8 Encapsulation

Encapsulation involves coating the oil droplets with materials to create small capsules, preventing their dispersion and facilitating their removal from the water [28]. The encapsulation process encapsulates the oil droplets within a protective shell, confining them and preventing their interaction with the surrounding environment. Here are key aspects of encapsulation:

I. **Encapsulating Materials**: Encapsulating materials can include natural or synthetic polymers, surfactants, or emulsifiers. These materials form a coating around the oil droplets, effectively trapping them within the capsules [29].
II. **Application Process**: Encapsulation can be achieved by adding the encapsulating materials directly to the oil spill or by creating emulsions or dispersions containing the encapsulating agents. These mixtures are then applied to the oil spill, allowing the encapsulating materials to coat the oil droplets [24].
III. **Benefits and Considerations**: Encapsulation helps prevent the oil from spreading, minimizing its contact with the water column and reducing its potential impact on aquatic life and ecosystems. The encapsulated oil can be subsequently collected using various methods such as skimming, sorbents, or physical separation techniques. However, the long-term stability and potential release of the encapsulated oil under certain conditions should be considered during the cleanup process [23, 30].

Both solidification and encapsulation techniques offer advantages in managing and mitigating oil spills. They can be particularly useful in situations where other recovery methods are limited or ineffective. However, the selection and application of these techniques should consider the specific characteristics of the spilled oil, environmental factors, and potential impacts on the ecosystem. Integration with other oil spill response measures, such as containment booms, skimmers, or sorbents, is often necessary to achieve comprehensive and effective cleanup and remediation. In situ burning and dispersants are two techniques commonly employed in oil spill response to mitigate the environmental impact of spills [23, 24, 29]. These techniques are utilized to control and reduce the spread of spilled oil, both on the water surface and below the surface. Let us explore in situ burning and dispersants in more detail.

## 2.9 In Situ Burning

In situ burning involves the controlled burning of spilled oil directly on the water surface [12, 31] shown in Fig. 2. This technique can be effective for treating large volumes of floating oil and has several key considerations:

**Fig. 2** In situ burning of an oil slick, or part of a slick, before it reaches the coast [7]

- In situ burning requires specific weather and environmental conditions, including suitable wind speeds, minimal wave activity, and appropriate air and water temperatures. These conditions help facilitate the combustion process and minimize smoke and residue.
- Prior to burning, booms are typically used to corral or contain the oil in a specific area. This helps concentrate the oil for efficient burning and prevents the spread of the fire. Fire-resistant booms are often used to provide additional safety during the burning operation.
- Ignition devices, such as flare guns or igniters, are used to initiate the controlled burn. The fire is managed by trained personnel who monitor the burn area, adjust the burn rate, and ensure the safety of the operation. The burning process converts the floating oil into heat, smoke, and residue, reducing the overall volume of oil and minimizing its environmental impact.
- In situ burning can produce smoke, emissions, and residue, which may have environmental impacts. Therefore, careful consideration must be given to air quality, potential effects on wildlife, and the surrounding ecosystem. Environmental assessments and permits are often required before implementing in situ burning operations.

### 2.9.1 Dispersants

Dispersants are chemical agents used to break down and disperse oil into smaller droplets, enhancing its natural dispersion in the water column. Dispersants facilitate the rapid dilution and biodegradation of the oil, reducing its impact on the water surface and allowing microbial degradation processes to take place [9, 11, 12, 32, 33]. Here are some key aspects of dispersant use:

- Dispersants are typically applied using specialized equipment, such as aerial sprayers or vessel-mounted spray systems. The dispersant is sprayed directly onto the oil slick, causing the oil to break up into smaller droplets that disperse throughout the water column.
- Dispersant application is most effective when applied early in the spill response, ideally within the first few hours to days after the spill occurs. Timing is crucial to maximize the dispersant's effectiveness in breaking down the oil. The effectiveness of dispersants can vary depending on factors such as oil type, weather conditions, and water temperature.
- While dispersants aid in the natural breakdown of oil, their use raises environmental concerns. Dispersants may have potential toxic effects on aquatic organisms and ecosystems. Therefore, their application should be carefully evaluated, considering factors such as the proximity of sensitive habitats, species at risk, and the potential for long-term ecological impacts. Regulatory requirements and environmental assessments are crucial before deploying dispersants.

In situ burning and dispersants are valuable tools in oil spill response, but their use should be carefully evaluated based on the specific spill characteristics, environmental considerations, and regulatory requirements. Integration with other response measures, such as containment, recovery, and shoreline cleanup, is often necessary to ensure a comprehensive and effective oil spill response strategy.

Chemical dispersants are specialized substances used in oil spill treatment to break down and disperse spilled oil into smaller droplets, promoting its rapid dilution and natural dispersion in the water column. The purpose of dispersants is to minimize the impact of oil spills on the water surface, shoreline, and sensitive habitats by facilitating the biodegradation and removal of the oil.

## 3 Mechanisms of Action

### *3.1 Dispersants Work Through Two Primary Mechanisms*

I. **Emulsification**: Dispersants contain surfactants, which are active agents that reduce the interfacial tension between oil and water. When dispersants are applied to the oil slick, the surfactants help to emulsify the oil into tiny droplets, forming a stable oil-in-water emulsion. This emulsification process increases the surface area of the oil, making it more susceptible to natural processes such as wave action, microbial degradation, and dispersion [28].

II. **Dispersion**: The emulsified oil droplets formed by dispersants disperse throughout the water column due to wave and current action. This dispersion spreads the oil over a larger volume, reducing its concentration and facilitating its dilution. It enhances the exposure of the oil to natural processes and microorganisms, accelerating its biodegradation and reducing its potential impact on sensitive environments [33].

## *3.2 Factors Influencing Dispersant Effectiveness*

Several factors influence the effectiveness of dispersants in oil spill treatment:

I. **Oil Properties**: Dispersant effectiveness can vary depending on the type of oil spilled. Factors such as oil viscosity, composition, weathering, and the presence of emulsifiers can impact how readily the oil disperses. Lighter oils and those with lower viscosity generally disperse more readily than heavier or highly viscous oils [34].
II. **Water Conditions**: Environmental factors, including water temperature, salinity, turbulence, and wave action, can influence the dispersant's effectiveness. Warmer water temperatures and moderate turbulence enhance the dispersibility of the oil, while colder water or extreme wave conditions may limit the effectiveness of dispersants [10].
III. **Dispersant Application**: Proper application techniques and timing are crucial for dispersant effectiveness. The dispersant should be applied directly to the oil slick, targeting areas with high oil concentrations. Early application, ideally within the first few hours to days after the spill, increases the likelihood of successful dispersion [18, 35].

## *3.3 Controversies and Environmental Impacts*

The use of dispersants in oil spill response has been subject to controversy and debate due to potential environmental impacts. Key concerns include:

I. **Toxicity**: Dispersants can have toxic effects on marine organisms, particularly if used in high concentrations or in sensitive habitats. The surfactants in dispersants may cause harm to marine life, including fish, invertebrates, and plankton. The toxicity depends on the dispersant formulation, the specific oil being treated, and the exposure conditions [32].
II. **Long-Term Effects**: The long-term ecological consequences of dispersant use are still a subject of study and debate. While dispersants aid in the natural breakdown of oil, the effects of the dispersed oil on marine ecosystems, including potential sub-lethal impacts and bioaccumulation in the food chain, require careful consideration [9, 14, 36, 37].
III. **Trade-Offs**: There is a trade-off between the immediate benefits of dispersant use, such as reducing oil slick size and preventing shoreline impacts, and the potential long-term environmental impacts. Balancing these trade-offs and

considering alternative response strategies is crucial in decision-making during oil spill incidents [38].

Regulatory frameworks and guidelines exist to govern the use of dispersants, ensuring their application is carefully assessed, authorized, and monitored. Environmental assessments, site-specific evaluations, and close collaboration with regulatory bodies are essential to minimize the potential environmental impacts of dispersant use during oil spill response [18]. Overall, while dispersants can be effective in promoting the natural dispersion and biodegradation of spilled oil, their use requires careful evaluation, considering the specific spill characteristics, environmental factors, and potential impacts on marine ecosystems [8, 34, 39]. Integration with other response techniques and regular reassessment of response strategies are necessary to ensure a balanced and environmentally responsible approach to oil spill treatment.

## 4 Remote Sensing

Remote sensing technologies play a crucial role in oil spill detection and monitoring by providing valuable information about the extent, location, and characteristics of spills. These techniques utilize various sensors, including satellites, aircraft, and unmanned aerial vehicles (UAVs), to capture imagery and data that help assess and respond to oil spills. Here is an overview of the use of remote sensing in oil spill response:

### *4.1 Use in Oil Spill Detection and Monitoring*

I. **Satellite Imagery**: Satellites equipped with optical sensors capture high-resolution images of the Earth's surface. These images can detect oil spills by identifying the spectral signature of the oil on the water surface. Satellite-based monitoring provides a broad coverage area, allowing for the detection of large spills and tracking their movement over time [40].

II. **Aerial Surveys**: Aircraft, such as planes or helicopters, equipped with specialized sensors, fly over oil spill areas to capture detailed imagery. These surveys enable close-range monitoring of oil slicks, providing valuable information about the spill's size, shape, and distribution. Aerial surveys are particularly useful for smaller spills or areas of complex coastal environments [34].

III. **Unmanned Aerial Vehicles (UAVs)**: UAVs, or drones, equipped with imaging sensors, offer a flexible and cost-effective means of collecting high-resolution imagery. They can be deployed rapidly and maneuvered to capture detailed data on oil spills, even in challenging or inaccessible areas. UAVs provide a valuable tool for real-time monitoring and decision-making during spill response operations [11].

## 4.2 Advantages of Remote Sensing Technologies

Remote sensing enables monitoring of large areas, both offshore and near shore, providing a comprehensive view of the extent of oil spills and their movement over time. Satellite imagery and aerial surveys provide near real-time information about oil spills, allowing for prompt response and mitigation efforts. Remote sensing techniques enable data collection without direct contact with the spill, minimizing the disturbance to the affected environment and reducing potential risks to personnel.

## 4.3 Limitations of Remote Sensing Technologies

1. **Weather Conditions**: Cloud cover, poor visibility, and adverse weather conditions can affect the quality and availability of satellite and aerial imagery, potentially limiting their effectiveness.
2. **Detection Limits**: Remote sensing techniques may struggle to detect smaller or subsurface spills, as they rely on the detection of oil's spectral signature on the water surface.
3. **Interpretation Challenges**: Interpreting remote sensing data requires expertise and can be subject to interpretation errors, particularly when distinguishing between natural phenomena and oil slicks.

## 4.4 Recent Advancements and Future Prospects

Recent advancements in remote sensing technologies have expanded the capabilities for oil spill detection and monitoring. These include:

1. **Enhanced Resolution**: Improved spatial resolution of satellite imagery allows for more detailed monitoring and identification of smaller oil spills.
2. **Multi-Sensor Integration**: Integration of multiple sensors, such as optical, infrared, and radar, provides a more comprehensive understanding of oil spill characteristics, including thickness, composition, and potential impacts.
3. **Automated Detection Algorithms**: Development of advanced algorithms and machine learning techniques enables automated detection and classification of oil spills, reducing the reliance on manual interpretation.
4. **Integration with Other Technologies**: Combining remote sensing data with other technologies, such as artificial intelligence, robotics, and data analytics, holds promise for more efficient and effective oil spill response.

The future of remote sensing in oil spill response involves the continued advancement of sensor technologies, improved data processing and interpretation algorithms, and increased integration with other emerging technologies. This will enhance the accuracy, timeliness, and reliability of oil spill detection, monitoring, and response efforts, contributing to more effective environmental protection and mitigation strategies.

## *4.5 Integrated Response Strategies*

An integrated approach is crucial in oil spill treatment to effectively mitigate the impacts of spills and facilitate a coordinated response effort. It involves the collaboration and coordination of various stakeholders, including government agencies, industry, response organizations, local communities, and environmental groups [11, 41]. Here is why an integrated approach is important and how coordination and collaboration among stakeholders contribute to successful oil spill response:-

### 4.5.1 Importance of an Integrated Approach

I. **Comprehensive Response**: Oil spill response involves multiple aspects, such as containment, recovery, shoreline protection, wildlife rehabilitation, and environmental monitoring. An integrated approach ensures that all necessary components are considered and implemented cohesively, maximizing the effectiveness of the response [18].

II. **Resource Optimization**: By coordinating efforts and sharing resources, an integrated approach optimizes the use of personnel, equipment, and expertise. It avoids duplication of efforts and minimizes response time, resulting in a more efficient and cost-effective response [17].

III. **Stakeholder Engagement**: An integrated approach facilitates the involvement of various stakeholders, ensuring that their perspectives, concerns, and expertise are considered in decision-making processes. It promotes transparency, trust, and public confidence in the response efforts [37].

### 4.5.2 Coordination and Collaboration Among Stakeholders

I. **Government Agencies**: Government agencies, such as environmental agencies, coast guards, and regulatory bodies, play a central role in coordinating and overseeing oil spill response. They establish frameworks, guidelines, and regulations, and facilitate collaboration among other stakeholders.

II. **Industry and Operators**: Oil companies and operators are responsible for spill prevention measures, emergency response planning, and providing resources and expertise during spill incidents. Collaboration between industry and response organizations is vital to ensure an effective and timely response.

III. **Response Organizations**: Specialized response organizations, often funded by the industry or government, provide expertise, equipment, and trained personnel for oil spill response. They work in close coordination with other stakeholders to implement response strategies and mobilize resources.

IV. **Local Communities and Indigenous Groups**: Involving local communities and indigenous groups in the response process is crucial. Their knowledge of the local environment, cultural sensitivities, and direct impacts helps shape response strategies and ensure effective communication and engagement.

### 4.5.3 Examples of Successful Integrated Response Strategies

I. **Deepwater Horizon Oil Spill**: The response to the Deepwater Horizon oil spill in 2010 demonstrated the importance of an integrated approach. Multiple agencies, industry, response organizations, and research institutions collaborated to address various aspects of the spill, including containment, dispersant application, shoreline protection, and wildlife rescue. Lessons learned from this event led to improvements in spill response planning and regulations.

II. **Baltic Sea Response**: The Baltic Sea is prone to oil spills due to heavy maritime traffic. In response, countries in the region, including Sweden, Finland, and Estonia, have established joint response plans, shared resources, and conducted regular joint exercises. This integrated approach ensures a coordinated response across national boundaries, improving readiness and effectiveness [42].

### 4.5.4 Lessons Learned

I. **Early Coordination**: Early coordination among stakeholders, even before an oil spill occurs, is crucial for preparedness and response. Establishing partnerships, sharing information, and conducting joint exercises enhance readiness and effectiveness.

II. **Communication and Information Sharing**: Effective communication channels and information sharing mechanisms are essential for seamless coordination. Timely and accurate information exchange among stakeholders helps align response efforts and decision-making.

III. **Continuous Learning and Adaptation**: Oil spill response strategies should be regularly reviewed, evaluated, and updated based on lessons learned from past incidents, new technologies, and evolving best practices. Continuous learning and adaptation contribute to ongoing improvements in oil spill response.

An integrated approach in oil spill treatment ensures a coordinated, efficient, and effective response. By leveraging the expertise, resources, and perspectives of various stakeholders, we can minimize the environmental impact of spills and protect affected ecosystems and communities.

## 4.6 Best Practices and Lessons Learned

Effective oil spill treatment relies on the implementation of best practices that encompass preparedness, response planning, training, and continuous improvement. Here are key best practices and lessons learned in oil spill treatment:

### 4.6.1 Training, Preparedness, and Response Planning

I. **Training Programs**: Implement comprehensive training programs for personnel involved in oil spill response, including responders, industry employees, and government agencies. Training should cover spill response techniques, equipment operation, safety protocols, and environmental considerations [30].

II. **Preparedness and Response Plans**: Develop and regularly update spill response plans that outline roles, responsibilities, and procedures for all stakeholders involved in oil spill treatment. Plans should address potential scenarios, coordination mechanisms, resource mobilization, and communication strategies [43].

III. **Exercises and Drills**: Conduct regular exercises and drills to test and enhance the effectiveness of spill response plans. These simulations provide opportunities to identify gaps, improve coordination, and evaluate the readiness of personnel and equipment [44, 45].

### 4.6.2 Early Response and Rapid Deployment

I. **Early Detection and Reporting**: Establish systems for early detection and reporting of oil spills to facilitate a prompt response. Encourage reporting mechanisms from various sources, including the public, industry, and surveillance technologies.

II. **Rapid Deployment**: Ensure a rapid and coordinated deployment of response resources to minimize the spread and impact of oil spills. Pre-position equipment and establish mechanisms for quick mobilization of trained personnel, containment booms, skimmers, dispersants, and other necessary resources.

### 4.6.3 Environmental Considerations

I. **Sensitive Area Mapping**: Identify and map sensitive ecosystems, habitats, and cultural or archaeological sites in the vicinity of potential spill areas. Incorporate this information into spill response plans to prioritize protection efforts and minimize ecological damage [35].

II. **Shoreline Protection**: Develop strategies and techniques for shoreline protection to prevent or mitigate oil from reaching coastal areas. This may include deploying booms, constructing barriers, and implementing shoreline cleanup methods specific to the affected environment [46].

### 4.6.4 Continuous Improvement and Lessons Learned

I. **Post-Incident Evaluation**: Conduct thorough post-incident evaluations to identify strengths, weaknesses, and areas for improvement in spill response efforts. Assess the effectiveness of strategies, technologies, and coordination mechanisms employed during the response.
II. **Incorporation of Lessons Learned**: Incorporate lessons learned from past spills, research, and advancements in technology into spill response planning. Regularly update response plans to reflect the latest knowledge and best practices.
III. **Research and Development**: Encourage research and development efforts to advance technologies, techniques, and approaches for oil spill treatment. Foster collaborations between academia, industry, and government to address knowledge gaps and enhance response capabilities.
IV. **International Cooperation**: Foster international cooperation and information exchange on oil spill response best practices, technologies, and lessons learned. Collaborate with other countries and organizations to enhance global readiness and response capacity.

In conclusion, effective oil spill treatment requires training, preparedness, and response planning. It involves early response, rapid deployment of resources, and consideration of environmental factors. Continuous improvement and the incorporation of lessons learned are vital for enhancing response capabilities and mitigating the impacts of oil spills. By implementing these best practices, we can better protect the environment, communities, and natural resources affected by such incidents.

## 5 Conclusion

In conclusion, the chapter provides an overview of mechanized solutions for oil spill treatment, covering various techniques and technologies employed in the containment, recovery, and treatment of oil spills. From skimmers and sorbents to mechanical separation and solidification techniques, the chapter explores the diverse range of approaches available for addressing oil spills. It further discusses the importance of an integrated approach that combines these techniques to maximize effectiveness and minimize environmental impacts. The chapter highlights the significance of remote sensing and robotic technologies in oil spill response, including the use of satellite imagery, aerial surveys, and other remote sensing techniques for detection and monitoring. It acknowledges the advantages and limitations of these technologies and emphasizes recent advancements and future prospects in the field, such as enhanced resolution, multi-sensor integration, and automated detection algorithms. Furthermore, the chapter emphasizes the role of chemical dispersants in oil spill treatment, including their purpose, mechanisms of action, and factors influencing effectiveness. It addresses the controversies and potential environmental impacts associated with dispersant use, emphasizing the need for careful evaluation and integration with other response techniques [43, 44].

The chapter concludes by discussing the importance of an integrated approach in oil spill treatment, highlighting the coordination and collaboration among different stakeholders. It emphasizes the need for comprehensive response strategies, stakeholder engagement, and optimized resource utilization. Examples of successful integrated response strategies are provided, such as the Deepwater Horizon oil spill and collaborative efforts in the Baltic Sea region. Lessons learned from these incidents emphasize the significance of early coordination, effective communication, continuous learning, and adaptation. Finally, the chapter summarizes key best practices for effective oil spill treatment, including training, preparedness, and response planning. It emphasizes the importance of early response, rapid deployment, and consideration of environmental factors. The chapter concludes by stressing the need for continuous improvement and the incorporation of lessons learned to enhance response capabilities and mitigate the impacts of oil spills.

Overall, the chapter delivers a comprehensive understanding of mechanized solutions for oil spill treatment, emphasizing the importance of an integrated approach, the advancements in remote sensing and robotic technologies, the role of chemical dispersants, and the implementation of best practices. By adopting these approaches, stakeholders can work collaboratively to protect the environment, minimize the damage caused by oil spills, and ensure effective and efficient oil spill response.

## References


1. Das N, Chandran P (2011) Microbial degradation of petroleum hydrocarbon contaminants: an overview. Biotechnol Res Int 2011:1–13
2. Brakstad OG, Lewis A, Beegle-Krause CJ (2018) A critical review of marine snow in the context of oil spills and oil spill dispersant treatment with focus on the Deepwater Horizon oil spill. Mar Pollut Bull 135:346–356
3. Ifelebuegu AO, Ukpebor JE, Ahukannah AU, Nnadi EO, Theophilus SC (2017) Environmental effects of crude oil spill on the physicochemical and hydrobiological characteristics of the Nun River, Niger Delta. Environ Monit Assess 189(4)
4. Praba Karana C, Rengasamy RS, Das D (2011) Oil spill cleanup by structured fibre assembly. Indian J Fibre Text Res 36(2):190–200
5. Najar IA, Ahmadi RB, Jamian MAH, Hamza HB, Ahmad A, Sin CH (2022) Site-specific ground response analysis using the geotechnical dataset in moderate seismicity region. Int J Mech 16(1):37–45
6. Ahmadi R, Najar IA, Abdullahi AF, Norazzlina M, Hamza H, Najar NA (2020) Computational investigation of soil liquefaction susceptibility based on standard penetration test value of Miri District (Sarawak, Malaysia). Int J Adv Sci Technol. 29(7):2735–2748
7. Obi EO, Kamgba FA, Obi DA (2014) Techniques of oil spill response in the sea. IOSR J Appl Phys 6(1):36–41

8. Dave D, Ghaly AE (2011) Remediation technologies for marine oil spills: a critical review and comparative analysis. Am J Environ Sci 7(5):424–440
9. Hoang AT, Nguyen XP, Duong XQ, Huynh TT (2021) Sorbent-based devices for the removal of spilled oil from water: a review. Environ Sci Pollut Res 28(23):28876–28910
10. Ventikos NP, Vergetis E, Psaraftis HN, Triantafyllou G (2004) A high-level synthesis of oil spill response equipment and countermeasures. J Hazard Mater 107(1–2):51–58
11. Doshi B, Sillanpää M, Kalliola S (2018) A review of bio-based materials for oil spill treatment. Water Res 135:262–277
12. Tuan Hoang A, Viet Pham V, Nam Nguyen D (2018) A report of oil spill recovery technologies. Int J Appl Eng Res 13(7):4915–4928. Available from: http://www.ripublication.com
13. Suni S, Kosunen AL, Hautala M, Pasila A, Romantschuk M (2004) Use of a by-product of peat excavation, cotton grass fibre, as a sorbent for oil-spills. Mar Pollut Bull 49(11–12):916–921
14. Liu H, Geng B, Chen Y, Wang H (2017) Review on the aerogel-type oil sorbents derived from nanocellulose. ACS Sustain Chem Eng 5(1):49–66
15. Tijani MM, Aqsha A, Mahinpey N (2016) Development of oil-spill sorbent from straw biomass waste: experiments and modeling studies. J Environ Manage 171:166–176. Available from: https://doi.org/10.1016/j.jenvman.2016.02.010
16. Wahi R, Chuah LA, Choong TSY, Ngaini Z, Nourouzi MM (2013) Oil removal from aqueous state by natural fibrous sorbent: an overview. Sep Purif Technol 113:51–63. Available from: https://doi.org/10.1016/j.seppur.2013.04.015
17. Zhong Z, You F (2011) Oil spill response planning with consideration of physicochemical evolution of the oil slick: a multiobjective optimization approach. Comput Chem Eng 35(8):1614–1630. Available from: https://doi.org/10.1016/j.compchemeng.2011.01.009
18. Barron MG (2012) Ecological impacts of the Deepwater horizon oil spill: implications for immunotoxicity. Toxicol Pathol 40(2):315–320
19. Idris J, Eyu GD, Mansor AM, Ahmad Z, Chukwuekezie CS (2014) A preliminary study of biodegradable waste as sorbent material for oil-spill cleanup. Sci World J 2014
20. Choi HM (1996) Needlepunched cotton nonwovens and other natural fibers as oil cleanup sorbents. J Environ Sci Heal Part A Toxic/Hazardous Subst Environ Eng 31(6):1441–1457
21. Ifelebuegu AO, Johnson A (2017) Nonconventional low-cost cellulose- and keratin-based biopolymeric sorbents for oil/water separation and spill cleanup: a review. Crit Rev Environ Sci Technol 47(11):964–1001
22. Inagaki M, Kawahara A, Nishi Y, Iwashita N (2002) Heavy oil sorption and recovery by using carbon fiber felts. Carbon N Y 40(9):1487–1492
23. Bakry AM, Abbas S, Ali B, Majeed H, Abouelwafa MY, Mousa A et al (2016) Microencapsulation of oils: a comprehensive review of benefits, techniques, and applications. Compr Rev Food Sci Food Saf 15(1):143–182
24. Alsmaeil AW, Hammami MA, Enotiadis A, Kanj MY, Giannelis EP (2021) Encapsulation of an anionic surfactant into hollow spherical nanosized capsules: size control, slow release, and potential use for enhanced oil recovery applications and environmental remediation. ACS Omega 6(8):5689–5697
25. ShamsiJazeyi H, Miller CA, Wong MS, Tour JM, Verduzco R (2014) Polymer-coated nanoparticles for enhanced oil recovery. J Appl Polym Sci 131(15):1–13
26. Motta FL, Stoyanov SR, Soares JBP (2018) Application of solidifiers for oil spill containment: a review. Chemosphere 194:837–846
27. Sundaravadivelu D, Suidan MT, Venosa AD, Rosales PI, Campo-Moreno P, Conmy RN (2016) Development of a testing protocol for oil solidifier effectiveness evaluation. Clean Technol Environ Policy 18(4):1141–1150
28. Liu H, Wang C, Zou S, Wei Z, Tong Z (2012) Simple, reversible emulsion system switched by pH on the basis of chitosan without any hydrophobic modification. Langmuir 28(30):11017–11024
29. Orlov MV (2021) Materials microencapsulation applications in oil drilling and production. J Phys Conf Ser 1942(1)

30. Halonen J, Lanki A (2019) Efficiency of maritime simulator training in oil spill response competence development. Trans Nav 13(1):199–204
31. Evans DD, Mulholland GW, Baum HR, Walton WD, McGrattan KB (2001) In situ burning of oil spills. J Res Natl Inst Stand Technol 106(1):231–278
32. Paul JH, Hollander D, Coble P, Daly KL, Murasko S, English D et al (2013) Toxicity and mutagenicity of gulf of Mexico waters during and after the Deepwater Horizon oil spill. Environ Sci Technol 47(17):9651–9659
33. Tansel B, Pascual B (2011) Removal of emulsified fuel oils from brackish and pond water by dissolved air flotation with and without polyelectrolyte use: Pilot-scale investigation for estuarine and near shore applications. Chemosphere 85(7):1182–1186. Available from: https://doi.org/10.1016/j.chemosphere.2011.07.006
34. Beyer J, Trannum HC, Bakke T, Hodson PV, Collier TK (2016) Environmental effects of the Deepwater Horizon oil spill: a review. Mar Pollut Bull 110(1):28–51. Available from: https://doi.org/10.1016/j.marpolbul.2016.06.027
35. Lehr WE (1974) Containment and recovery devices for oil spill cleanup operations. JPT, J Pet Technol 26:375–380
36. Sabir S (2015) Approach of cost-effective adsorbents for oil removal from oily water. Crit Rev Environ Sci Technol 45(17):1916–1945
37. Chang SE, Stone J, Demes K, Piscitelli M (2014) Consequences of oil spills: a review and framework for informing planning. Ecol Soc 19(2)
38. Osborne OE, Willie MMC, O'Hara PD (2022) The effects of oil spill dispersant use on marine birds: a review of scientific literature and identification of information gaps. Environ Rev 31(2):243–255
39. Annunciado TR, Sydenstricker THD, Amico SC (2005) Experimental investigation of various vegetable fibers as sorbent materials for oil spills. Mar Pollut Bull 50(11):1340–1346
40. Najar IA, Ahmadi R, Khalik YKA, Mohamad NZ, Jamian MAH, Najar NA (2022) A framework of systematic land use vulnerability modeling based on seismic microzonation: a case study of Miri District of Sarawak, Malaysia. Int J Des Nat Ecodyn 17(5):669–677
41. Abdel-Moghny T, Keshawy M (2014) An overview on the treatment of oil spill using sorbent materials: synthetic and natural oil sorbent. LAP Lambert Academic Publishing, pp 1–35
42. Brusendorff AC, Korpinen S, Meski L, Stankiewicz M (2014) HELCOM actions to eliminate illegal and accidental oil pollution from ships in the Baltic Sea, vol 27. Handbook of Environmental Chemistry, pp 15–40
43. Najar IA, Ahmadi R, Amuda AG, Mourad R, Bendary NE, Ismail I, Bakar NA, Tang S (2025) Advancing soil-structure interaction (SSI): a comprehensive review of current practices, challenges, and future directions. J Infrastruct Preserv Resil 6(1):5
44. Skeie GM, Kristoffersen IS, Engen F, Sævik PM, Lura H, Johnsen GH (2021) Integration of lessons learned from 30 years of oil on water exercises, testing and training in an improved industry oil spill contingency framework. Int Oil Spill Conf Proc 2021(1)
45. Greenberg B, Voevodsky P, Gralla E (2016) A capabilities-based framework for disaster response exercise design and evaluation: findings from oil spill response exercises. J Homel Secur Emerg Manag 13(4):1–18
46. Etkin DS (2004) Twenty-year trend analysis of oil spills in EPA jurisdiction. In: Proceedings of the 5th biennial freshwater spills symposium, pp 1–15. Available from: http://environmental-research.com/erc_papers/ERC_paper_15.pdf

# Impacts of Oil Spill on Geotechnical Properties of Soil

**Imtiyaz Akbar Najar, Raudhah Ahmadi, Showkat Ahmad Bhawani, Murtala Namakka, and Nur Hisyam Ramli**

**Abstract** The escalation of environmental pollution in the world is attributed to the recent increase in crude oil spills, mainly due to the increase in crude oil production and the activities of individuals involved in crude oil theft. In oil and gas, the likelihood of soil pollution from petroleum products soars as oil exploration, production, and transportation of substantial oil volumes via pipelines across vast distances continue to dilate. Recent findings divulge a plethora of instances of oil spills occurring in various locations, attributed to damage within the infrastructure of the oil industry. These spills have resulted in soil contamination, giving rise to significant environmental risks and degradation of both the soil and its engineering characteristics. Herein we explore the changes in geotechnical properties of soil resulting from oil contamination. Beyond understanding the detrimental effects, the chapter explores numerous treatment methodologies, both established and emerging, offering a roadmap for restoring the geotechnical integrity of oil-contaminated soil. The advantages of employing geotechnical approaches, including sustainability, cost-effectiveness, and long-term efficacy, are highlighted, looking ahead, we identify key areas emphasizing the need for long-term monitoring, exploration of emerging contaminants and technologies like nanotechnology, for the development of scalable and cost-effective solutions.

**Keywords** Geotechnical · Oil spill · Pollution · Soil · Contamination

---

I. A. Najar (✉) · R. Ahmadi · M. Namakka · N. H. Ramli
Faculty of Engineering, Universiti Malaysia Sarawak, Kota Samarahan, Sarawak, Malaysia
e-mail: imtiyaznajar999@gmail.com

S. A. Bhawani
Faculty of Resource Science and Technology, Universiti Malaysia Sarawak, Kota Samarahan, Sarawak, Malaysia

© The Author(s), under exclusive license to Springer Nature Singapore Pte Ltd. 2025 
S. A. Bhawani et al. (eds.), *Biobased Materials and Their Composites for Oil Spill Treatment*, Composites Science and Technology,
https://doi.org/10.1007/978-981-96-6049-0_3

# 1 Introduction

An oil spill refers to the release or discharge of liquid petroleum hydrocarbons, commonly known as oil, into the environment [1]. These spills can occur in various locations, including land, water bodies (such as oceans, seas, rivers, or lakes), or even underground. Oil spills are typically the result of human activities, accidents, or natural disasters, and they can have detrimental effects on ecosystems, wildlife, and human communities [2]. The most common type of oil spills occurs in marine environments, where crude oil or refined petroleum products are released into the water [1–3]. These spills may be the consequence of incidents involving offshore drilling platforms, pipelines, oil tankers, or other oil-related infrastructure. Additionally, spills on land can occur during transportation, storage, or industrial processes. The consequences of oil spills are extensive and include environmental pollution, damage to aquatic and terrestrial ecosystems, harm to wildlife, and economic losses for industries such as fishing and tourism [4]. Cleanup efforts for oil spills involve various techniques and technologies [5], including containment booms, skimmers, dispersants, and in-situ burning [2–4]. Preventing oil spills and developing effective response strategies are crucial for minimizing the environmental impact and protecting both natural ecosystems and human activities that rely on clean water and land [3–5].

Consequentially, oil spills can have significant impacts on the geotechnical parameters of soil, influencing its mechanical and hydraulic properties [6]. The extent of these impacts depends on various factors such as the type and amount of oil spilled, the characteristics of the soil, and the environmental conditions [7]. Table 1 outlined some of the influence of oil spills on the performance and properties of the affected terrain.

It is crucial to note that the use of remediation methods, such as soil washing or bioremediation, can also influence geotechnical parameters [7, 8]. Moreso, some methods may further disturb the soil structure that the specific impact of an oil spill on soil will vary based on site-specific conditions [9]. Hence, proper assessment and monitoring are essential to understand the full extent of the effects and to implement appropriate remediation strategies [2–7]. Additionally, local regulations and guidelines may dictate the required actions for mitigating the impact of oil spills on soil quality.

An oil spill on soil can have profound effects on its geotechnical properties, introducing significant challenges for environmental and geotechnical engineers [5]. The spilled oil permeates the soil, leading to changes in its physical and mechanical characteristics. The hydrophobic nature of oil can alter the soil's porosity [10], reducing water infiltration and impeding natural drainage processes [11]. This, in turn, may affect the soil's shear strength, compaction, and overall stability. The oil's interaction with soil particles can lead to increased soil hydrophobicity, causing shifts in aggregate size distribution [11–19]. Furthermore, the contamination may result in changes to the Atterberg limits, such as an increase in liquid limit [6], plastic limit [4, 28], and plasticity index [13]. These alterations in geotechnical properties pose challenges

**Table 1** Effects and mechanisms of oil spill on geotechnical properties if oil contaminated soil

| Oil spill affected material type | Mechanism | Effect on soil geotechnical properties |
|---|---|---|
| Oil contaminated soil | Coat soil particles, making them hydrophobic | Reduces water infiltration and affects the soil's ability to retain moisture |
| Oil contaminated soil | Clog soil pores | Reducing permeability and altering the flow of water through the soil. This can lead to increased runoff and erosion |
| Oil contaminated soil | Reduction in shear strength | Instability of slopes and foundations |
| Oil contaminated soil | Alter the compaction characteristics of soil | Affecting oil density and porosity |
| Contaminated soil | Oil components-to-soil minerals interactions, imbalance in soil chemical compositions | Alters the soil's mineralogy and potentially affects its engineering properties |
| Oil contaminated soil | Microbial decomposition | Production of byproducts which alter the soil structure and composition |
| Oil contaminated soil | Root penetration and growth | Instability of vegetation and increasing the risk of erosion |

for construction projects in affected areas, necessitating a thorough understanding of the specific impacts of oil contamination on soil behavior [14]. Remediation efforts to restore the geotechnical integrity of the soil often require a multidisciplinary approach, combining environmental science, engineering, exploration of the potential utilization of nanotechnology [32], and regulatory compliance.

## 2 State-Of-The-Art and Oil Spill Sources

The primary sources of surface contamination by crude oil include leakages from pipelines, oil wells, underground storage tanks, and spills from accidents like the Gulf War oil spills and the Valdez tanker incident. Notably, Saudi Aramco (2005) reported that over half the world's crude oil comes from the Arabian Gulf, where a staggering 550 spills totalling 14,000 barrels occurred between 1995 and 1999, followed by another 11,000 barrels between 2000 and 2003 [5]. This oil contamination poses not only environmental threats but also geotechnical challenges due to its impact on soil properties, as documented by numerous studies over the past decade [7–12].

Despite its detrimental environmental consequences [15], petroleum, a complex organic mixture of aromatic hydrocarbons, asphaltenes, and heavy metals [13], extracted through underground exploration [14], continues to be a globally increasing resource [15]. Petroleum production primarily encompasses exploration and extraction, refining and marketing, and transportation [15, 16]. However, mismanagement of both petroleum production and its derivatives can lead to significant environmental pollution, particularly through oil spills [15]. This poses a major threat in oil-producing regions like Nigeria, where vandalism and crude oil theft often contribute to a high incidence of spills [14–20]. The resulting crude oil pollution negatively impacts the physical, mechanical, and biochemical properties of surrounding soil, potentially affecting nearby structures like building foundations, road pavements, and earth-retaining structures [21, 22]. Interestingly, Khamehchiyan et al. [23] suggest that with lower oil contamination (<16%), the likelihood of natural drainage from the soil diminishes significantly.

kinwumi et al. [24] and Al-Obaidy et al. [25] highlight the multifaceted impact of oil spills on soil properties, emphasizing both alterations in engineering behavior and environmental ramifications. Spilled oil elevates hydrocarbons as well as toxic metals in soil [25], triggering substantial modifications in its chemical, mechanical, physical, and biological characteristics [9, 16, 26]. The extent of these alterations depends on the oil volume and type, as well as the soil's inherent features [17, 24]. Significant consequences include reductions in Atterberg limits, compaction, permeability, angle of internal friction, cohesion, and consolidation properties [17, 24]. Oil disrupts internal structure, decreasing frictional resistance and influencing compression characteristics, particularly in sandy soils [2, 9, 27]. Furthermore, oil pollution alters particle surface texture and promotes the formation of atypical pores within the soil matrix [4], further compromising its engineering behavior and contributing to environmental concerns.

## 3 Influence of Oil Spill on Soil Geotechnical Properties

Hydrocarbon-contaminated soils typically exhibit a denser, flocculated structure with increased Edge-to-Face contacts amongst particles [17, 18]. In contrast, uncontaminated fines tend to display a more open and recognizable fabric characterized by prevalent Face-to-Face and Edge-to-Edge arrangements [28]. This shift in microstructure, as Izdebska-Mucha and Trzciński observed, transforms the soil's initially anisotropic pattern into a uniform matrix upon petroleum exposure [28]. The degree to which this transformation affects soil's geotechnical properties depends on multiple factors, including the volume and concentration of the contaminant, climatic conditions, soil's physical characteristics, vegetation cover, and its chemical composition [2, 17, 18].

Nudelman et al. [29] illuminate the crucial role of soil properties, such as permeability, adsorption rate, and partition coefficient, in dictating the rate at which oil modifies soil behavior. Evgin and Das [8] further utilize finite-element analysis to demonstrate how oil presence in soil amplifies settlement, particularly in footings. Notably, oil evaporation from the soil surface can alter the oil's physiochemical properties, facilitating restoration of the geotechnical characteristics of the affected soil [30]. This interplay between oil and soil highlights the dynamic nature of their interaction and its significant impact on geotechnical properties.

### 3.1 Impact on Atterberg Limits

In their studies on various geotechnical properties of soils, Tse and Eshiemomo, [31] and Evgin and Das, [8] observed significant impacts of oil contamination on the Atterberg limits and strength parameters of the soils. Khamehchiyan et al. [23] investigated the geotechnical properties of clayey and sandy soils contaminated with oil in Iran. Their findings indicated a general decline in the strength, permeability, maximum dry density (MDD), optimum water content (OWC), and Atterberg limits of the soil due to oil contamination. Youdeowei [21] concluded in their respective studies that soil liquid limits and plastic limits exhibited a notable decrease following oil pollution. According to Youdeowei [21], the liquid limits of the soils decreased by 8%, while the plastic limits and plasticity indices witnessed declines of 9% and 40%, respectively, after petroleum contamination. Additionally, Rahman et al. [32] and Rasheed et al. [33] investigated the impact of hydrocarbons on soil geotechnical properties. Their findings indicated that with an increase in the volume of oil in the soil, there was a corresponding reduction in the soil's liquid limit and plastic limit values. Contrarily, Akinwumi et al. [24], Al-Obaidy et al. [25], and Khosravi et al. [34] argued that when fine-grained soils contained a substantial amount of petroleum, specifically exceeding 12%, there was a notable decrease in the soils' plastic limits. Contrarily, they noted a significant rise in both the liquid limits and plasticity indices. According to their findings, these anomalies in the Atterberg's limits were linked to the particle size characteristics of the soils. The presence of oil molecules, which exhibit non-polarity, and the distribution of particle sizes in the soils were identified as key factors contributing to the observed reduction in both liquid and plastic limits of fine-grained soils following contamination. Fine-grained soils do not exhibit plasticity with non-polar fluids [35, 36].

### 3.2 Impact on Permeability and Compaction

Puri [9] and Yazdi and Teshnizi [37] demonstrate that petroleum contamination significantly alters soil behavior, with profound reductions in compaction, permeability, and internal friction angle observed as petroleum content increases [37]. Notably, while Nazir [38] reports diminished permeability in granular soils, clayey soils exhibit minimal change due to reduced electrostatic interactions between clay particles and the non-polar oil [30, 39]. This interplay between petroleum and soil type highlights the multifaceted nature of contamination, emphasizing the need for tailored remediation strategies based on soil composition. Petroleum's detrimental impact on soil engineering properties is evident in its ability to alter permeability and density. Ilojeje and Aniago [40] observed a shift from silty clay to clayey soil behavior, accompanied by an 8% permeability decrease with increasing oil content. Similarly, studies by [2, 23, 37, 41] confirmed a significant oil-induced decrease in soil density. Rehman et al. [20] shed further light on this phenomenon, demonstrating that oil acts as a lubricating agent and binder in clay soils, leading to higher Maximum Dry Density (MDD) at lower water content. This interplay between oil and soil highlights the substantial modifications petroleum can induce, compromising the critical engineering properties of soil. However, it is noteworthy that higher petroleum content (>16%) resulted in a reduction in the soil MDD.

Moreover, in their respective studies on the influence of petroleum on soil geotechnical properties, Akinwumi et al. [24], Safehian et al. [39], and Akpokodje and Uguru [42] observed decreases in the Maximum Dry Density (MDD), Optimum Water Content (OWC), and permeability following soil contamination. The reduction in soil permeability is explained by the clogging of soil voids resulting from petroleum impact, as highlighted by Akpokodje and Uguru [42]. Additionally, Srivastava and Pandey, [43] found that an increase (from 0 to 12%) in petroleum content in alluvial soils leads to a consistent decrease in both the Maximum Dry Density (MDD) and Optimum Moisture Content (OMC) of the soils. Contrastingly, Singh et al. [44] asserted that the Optimum Moisture Content (OMC) of soil samples exhibited an increase of nearly 45% with the rise in the volume of petroleum in the soil samples, ranging from 0 to 9%.

### 3.3 *Impact on Hydraulic Conductivity of Soil*

Corroborating findings from Shin and Das [11] and Rojas et al. [45], Chew and Lee [46] solidify the notion that oil contamination directly and steadily diminishes the hydraulic conductivity of soils [2, 11, 45, 46]. Rojas et al. [45] further elucidate this relationship by demonstrating a concurrent rise in kinematic viscosity alongside decreased hydraulic conductivity [45]. This observed reduction in permeability is attributed to the physical obstruction of soil pores by oil, impeding water flow [46]. This consistent evidence across varied soil types and oil concentrations [11, 45, 46] underscores the profound impact of oil spills on soil's fundamental hydraulic properties.

### 3.4 *Impact on Soil Consolidation*

Consolidation refers to the prolonged application of a static load to a substantial soil mass with the aim of minimizing voids within the soil structure [47]. Also, the influence of oil on kaolinitic soil samples, a non-linear decline in void ratio and compressibility of the soil samples was observed as the volume of oil increased. Khosravi et al. [34] observed a general reduction in the soil compression index (Cc) in their experimental work after simulating an oil spill. However, the soil swelling index (Cs) was noted to increase at a gradual pace as the volume of oil increased. According to Safehian et al. [39] and Salimnezhad et al. [48], oil pollution led to a significant increase in the free swelling rate of soils, while the swelling pressure decreased due to oil contamination. Additionally, Talukdar and Saikia [47], in their examination of the geotechnical properties of fine-grained soils, found that the soil compression index (Cc) tends to decrease with an increase in the volume of oil resulting from an oil spill. In contrast, Srivastava, and Pandey [43] reported in their research that the soil compression index (Cc) tends to increase as the volume of oil in the soil rises.

# 4 Oil Treatment Methodologies

Oil spill treatment involves a combination of mechanical, chemical, and biological methods to contain, recover, and mitigate the impact of spilled oil on the environment. The choice of methods depends on factors such as the type of oil, environmental conditions, and the scale of the spill. Common methods used for oil spill treatment are depicted in Fig. 1.

## *4.1 Treatment of Oil Spill on Soil Through Geotechnical Approaches*

The treatment of oil spills on soil through geotechnical approaches involves a combination of physical, chemical, and biological methods aimed at mitigating the environmental impact and restoring the geotechnical properties of the affected soil. Some key geotechnical strategies for oil spill treatment on soil are summarized in Table 2.

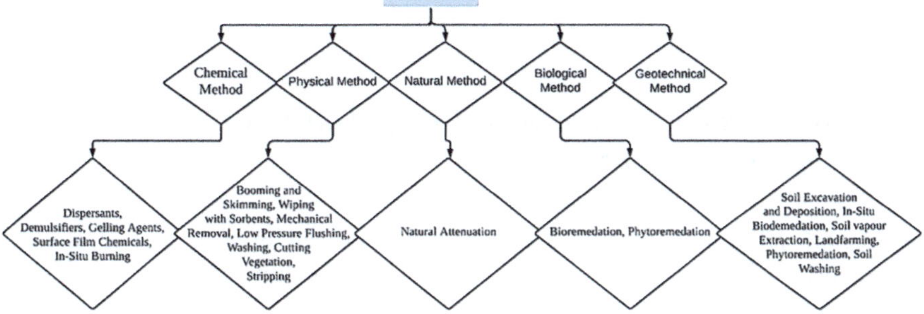


**Fig. 1** Methods of oil spill treatment

**Table 2** Oil contaminated soil treatment methods, applications, and mechanism

| Treatment methods | Mechanisms | Applications |
|---|---|---|
| Soil excavation and disposal | Remove and dispose of the contaminated soil to prevent further spreading of the oil | For heavily contaminated areas. Excavated soil may be substituted with clean soil |
| In-situ bioremediation | Stimulate the activity of indigenous microorganisms in the soil to naturally break down and degrade the oil. Bioremediation can be boosted by adding nutrients, oxygen, or microbial agents to accelerate the degradation process | Lightly contaminated soil |
| Soil vapor extraction | Use soil vapor extraction systems to remove volatile organic compounds from the soil. This technique involves the extraction of air and contaminants from the soil, followed by treatment and monitoring | Lower to mid-level oil contaminated soil |
| Landfarming | Encourage the natural degradation of oil in the soil by incorporating it into the top layer and providing optimal conditions for microbial activity | Lower-level contamination |
| Phytoremediation | Utilize plants with the ability to absorb and gather contaminants, a process known as phytoremediation. Certain plant species can enhance the degradation of oil in the soil | Mid-level contamination |
| Soil washing | Washing the soil with water or other washing agents and separating the contaminated components | Lower-level contamination |

However, effective oil spill treatment on soil through geotechnical approaches requires a site-specific assessment, considering factors such as soil type, contamination levels, and the surrounding environment. Integrating these geotechnical methods with environmental and regulatory considerations ensures a comprehensive and sustainable remediation process.

## *4.2 Advantages of Treatment of Oil Spill on Soil Through Geotechnical Approaches*

Treating oil spills on soil through a geotechnical approach offers several advantages, leveraging principles of soil mechanics and engineering to mitigate the environmental impact. Here are some key advantages:

- Geotechnical approaches often allow for in situ treatment, diminishing the need for excavation or removal of contaminated soil. This decreases disturbance to the natural environment and facilitates a more sustainable remediation process.
- Geotechnical methods aim to preserve the physical structure of the soil. Techniques such as soil vapor extraction and bioventing can be applied to remediate soil without causing significant disruption to the soil matrix, maintaining its integrity and fertility.
- Traditional excavation methods can lead to soil erosion and disruption of the natural landscape. Geotechnical methods, by contrast, can be designed to minimize such disturbances, preventing erosion and the potential for downstream impacts on water bodies.
- Geotechnical methods can enhance natural biodegradation processes by optimizing conditions for microbial movement in the soil. This includes adjusting factors such as moisture content, aeration, and nutrient levels to promote the growth of indigenous microorganisms that can break down the contaminants.
- Certain geotechnical techniques involve the use of alterations or barriers to immobilize contaminants in the soil. This prevents the migration of impurities to groundwater or other sensitive receptors, offering a long-term solution to contamination.
- In many cases, geotechnical approaches can be more cost-effective compared to extensive excavation and off-site disposal. Treating the contamination in place reduces transportation and disposal costs associated with excavated soil.
- The geotechnical approach, particularly in situ treatments, minimizes the disruption of the affected site. This is particularly advantageous in urban or ecologically sensitive areas where preserving the existing landscape is a priority.
- Geotechnical methods can be customized based on the specific characteristics of the contaminated site, allowing for a targeted and efficient remediation strategy. This adaptability is crucial for addressing diverse soil types, hydrogeological conditions, and contaminant compositions.
- Geotechnical methods are designed to meet environmental regulations and guidelines. Implementing a geotechnical approach ensures that the remediation process aligns with regulatory requirements, promoting responsible and sustainable practices.
- When appropriately designed and implemented, geotechnical methods can provide long-term effectiveness in treating oil-contaminated soil. This helps prevent the recurrence of contamination and promotes the restoration of the soil ecosystem.

In summary, the geotechnical approach to oil spill treatment on soil offers a suite of advantages, including in situ treatment, preservation of soil structure, cost-effectiveness, and tailored remediation strategies that make it a valuable and environmentally conscious option for addressing soil contamination.

## 5 Conclusion and Future Perspectives

In this Chapter we explored the intricate relationship between oil spills and the geotechnical properties of soil. We have delved into the multifaceted effects and mechanisms at play (outlined in Table 1), analyzed the impact of petroleum on soil characteristics, and examined several promising oil treatment methodologies (see Table 2). Notably, the advantages of employing geotechnical approaches for oil spill remediation have been highlighted, emphasizing their potential for efficient and sustainable restoration.

**Future Perspectives**:

Despite significant advancements in discerning and mitigating oil spill impacts on soil, several key areas assure further exploration:

- Long-term monitoring and assessment: Long-term monitoring of treated sites is crucial to assess the effectiveness and sustainability of remediation strategies. Developing robust predictive models to anticipate the long-term evolution of oil-contaminated soil will be essential for effective management.
- Emerging contaminants: With the increasing use of unconventional hydrocarbons and complex oil mixtures, research on their specific interactions with soil and the effectiveness of existing treatment methods is necessary.
- Nanotechnology: Exploring the potential of nanomaterials for targeted contaminant removal as revealed in the study of Namakka et al. [49], and investigating synergistic combinations of these technologies hold promising opportunities for enhanced efficacy and environmental footprint reduction.
- Cost-effective and scalable solutions: Developing cost-effective and scalable treatment solutions, while considering local soil conditions and resource availability, is critical for wider adoption of geotechnical approaches in real-world spill situations.
- Policy and regulatory frameworks: Establishing robust policies and regulatory frameworks that encourage preventive measures, prioritize geotechnical approaches, and promote responsible land management practices will play a crucial role in minimizing the long-term impacts of oil spills.

By tackling these future research directions, we can further enhance our understanding of oil-soil interactions, optimize existing remediation technologies, and develop innovative solutions.

## References

1. Saharan Y, Singh J, Goyat R, Umar A, Ibrahim AA, Akbar S, Baskoutas S (2023) Recent advances in soil cleanup technologies for oil spills: a systematic review. Water Air Soil Pollut 234(8):503

2. Najar IA, Ahmadi RB, Jamian MAH, Hamza H, Ahmad A, Sin CH (2022) Site-specific ground response analysis using the geotechnical dataset in moderate seismicity region. Int J Mech 16:37–45
3. Ahmadi R, Najar IA, Abdullahi AF, Sa'don NM, Hamza H, Najar NA (2020) Computational investigation of soil liquefaction susceptibility based on standard penetration test value of Miri District (Sarawak, Malaysia). Int J Adv Sci Technol 29(7):2735–2748
4. Haghsheno H, Arabani M (2022) Geotechnical properties of oil-polluted soil: a review. Environ Sci Pollut Res 29(22):32670–32701
5. Li P, Cai Q, Lin W, Chen B, Zhang B (2016) Offshore oil spill response practices and emerging challenges. Mar Pollut Bull 110(1):6–27
6. Al-Sanad HA, Ismael NF (1997) Aging effects on oil contaminated Kuwaiti sand. J Geotech Geoenviron Eng 123(3):290–293
7. Al-Sanad HA, Eid WK, Ismael NF (1995) Geotechnical properties of oil-contaminated Kuwaiti sand. J Geotech Eng 121(5):407–412
8. Evgin E, Das BM (1992) Mechanical behaviour of oil Akpokodje et al. 5 contaminated sand. In: Usmen MA, Acar YB (eds) Environmental geotechnology. Balkema Publishers, Rotterdam, Netherlands, pp 101–108
9. Puri VK (2000) Geotechnical aspects of oil-contaminated sands. Soil Sediment Contam 9(4):359–374
10. Puri VK, Das BM, Cook EE, Shin EC (1994) Geotechnical properties of crude oil contaminated sand. ASTM Spec Tech Publ 1221:75–88
11. Shin EC, Das BM (2000) Some physical properties of unsaturated oil-contaminated sand. In: Advances in unsaturated geotechnics, pp 142–152
12. Shin EC, Omar MT, Tahmaz AA, Das BM, Atalar C, De Mello LG (2002) Shear strength and hydraulic conductivity of oil-contaminated sand. In: Proceedings of the fourth international congress on environmental geotechnics, Rio de Janeiro, Brazil, vol 1. AA Balkema Publishers Lisse, pp 9–13
13. Moubasher HA, Hegazy AK, Mohamed NH, Moustafa YM, Kabiel HF, Hamad AA (2015) Phytoremediation of soils polluted with crude petroleum oil using *Bassia scoparia* and its associated rhizosphere microorganisms. Int Biodeterior Biodegradation 98:113–120
14. Uguru H, Udubra EA (2021) Optimizing the bioremediation of petroleum hydrocarbons contaminated soil. Int J Innov Environ Stud Res 9(3):47–53
15. Akpomrere OR, Uguru H (2020) Ecotoxicity effect of illegal refineries on the environment: a case study of Delta State, Nigeria. Int J Innov Agric Biol Res 8(2):40–49
16. Akpokodje OI, Uguru H (2019) Phytoremediation of petroleum products contaminated soil. Arch Curr Res Int 18(1):1–8
17. Akpomrere OR, Uguru H (2020) Spatial distribution of residual petroleum hydrocarbons in an oil spill site located at Isoko South LGA, Delta State, Nigeria. J Environ Waste Manage 7(1):312–317
18. Uguru H, Akpokodje OI, Esegbuyota D (2020) Remediation potency of charcoal block and sawdust in petroleum products contaminated soil. Trends Tech Sci Res 4(4):131–140
19. Udonne JD, Onwuma HO (2014) A study of effects of waste lubricating oil on the physical/chemical properties of soil and the possible remedies. J Petrol Gas Eng 5(1):9–14
20. Rehman H, Abduljauwad SN, Akram T (2007) Geotechnical behavior of oil-contaminated fine-grained soils. Electron J Geotech Eng 12:1–12
21. Youdeowei PO (2008) The effect of crude oil pollution and subsequent fire on the engineering properties of soils in the Niger Delta. Bull Eng Geol Env 67(1):119–121
22. Hong JH, Kim J, Choi OK, Cho KS, Ryu HW (2005) Characterization of a diesel-degrading bacterium, *Pseudomonas aeruginosa* IU5, isolated from oil-contaminated soil in Korea. World J Microbiol Biotechnol 21(3):381–384
23. Khamehchiyan M, Charkhabi AH, Tajik M (2007) Effects of crude oil contamination on geotechnical properties of clayey and sandy soils. Eng Geol 89(3–4):220–229
24. Akinwumi II, Diwa D, Obianigwe N (2014) Effects of crude oil contamination on the index properties, strength and permeability of lateritic clay. Int J Appl Sci Eng Res 3(4):1–10

25. Al-Obaidy N, Al-Shueli A, Sattar H, Majeed Z, Hamid NAH (2019) An experimental study on geotechnical and electrical properties of an oil-contaminated soil at thi-qar governorate/Iraq. Int Rev Civil Eng 10(3):148–154
26. Chang SE, Stone J, Demes K, Piscitelli M (2014) Consequences of oil spills: a review and framework for informing planning. Ecol Soc 19(2):26
27. Evgin E, Amor FB, Altaee A (1989) Effect of an oil spill on soil properties. In: Proceedings of the 8th international conference on offshore mechanics and Arctic engineering. American Society of Mechanical Engineering, Hague, p 715 [6 pp]
28. Izdebska-Mucha D, Wójcik E (2022) Shrinkage properties of natural and diesel oil-contaminated fine soils from central Poland. Quart J Eng Geol Hydrogeol 55(2):qjegh2021-006
29. Nudelman NS, Rios IS, Katusich O (2002) Fate of the oil residuals in Patagonian soils effects of the environmental exposure time. J Environ Risk Assess Rem 3:1–8
30. Ostovar M, Ghiassi R, Mehdizadeh MJ, Shariatmadari N (2020) Effects of crude oil on geotechnical specification of sandy soils. Soil Sediment Contam Int J 30(1):58–73
31. Tse AC, Eshiemomo AU (2016) Geotechnical properties of soils in a crude oil impacted site in the Niger Delta, Nigeria. IOSR J Appl Geol Geophys 4(2):69–76
32. Rahman ZA, Mohd UH, Taha R (2010) Influence of oil contamination on geotechnical properties of basaltic residual soil. Am J Appl Sci 7(7):954–961
33. Rasheed ZN, Ahmed FR, Jassim HM (2014) Effect of crude oil products on the geotechnical properties of soil. Witpress Ltd
34. Khosravi E, Ghasemzadeh H, Sabour MR, Yazdani H (2013) Geotechnical properties of gas oil-contaminated kaolinite. Eng Geol 166:11–16
35. Kaya A, Fang HY (2000) The effects of organic fluids on physicochemical parameters of fine-grained soils. Can Geotech J 37(5):943–950
36. Rajabi H, Sharifipour M (2019) Geotechnical properties of hydrocarbon-contaminated soils: a comprehensive review. Bull Eng Geol Env 78(5):3685–3717
37. Yazdi A, Sharifi Teshnizi E (2021) Effects of contamination with gasoline on engineering properties of fine-grained silty soils with an emphasis on the duration of exposure. SN Appl Sci 3(7):1–24
38. Nazir AK (2011) Effect of motor oil contamination on geotechnical properties of over consolidated clay. Alexandria Eng J 50(4):331–335
39. Safehian H, Rajabi AM, Ghasemzadeh H (2018) Effect of diesel-contamination on geotechnical properties of illite soil. Eng Geol 241:55–63
40. Iloeje AF, Aniago V (2016) Effect of crude oil on permeability properties of the soil. Int J Trend Sci Res Develop 1(1):39–43
41. Shah SJ, Shroff AV, Patel JV, Tiwari KC, Ramakrishnan D (2003) Stabilization of fuel oil contaminated soil—a case study. Geotech Geol Eng 21(4):415–427
42. Akpokodje OI, Uguru H (2019) Bioremediation of hydrocarbon contaminated soil: assessment of compost manure and organic soap. Trans Mach Learn Artif Intell 7(5):13–23
43. Srivastava RK, Pandey VD (1998) Geotechnical evaluation of oil contaminated soil. In: Sarsby RW (ed) The proceeding of CREEN 2 in the second international symposium on geotechnics related to the environment Krakow. Poland Thomas Telford, London
44. Singh SK, Srivastava RK, John S (2009) Studies on soil contamination due to used motor oil and its remediation. Can Geotech J 46(9):1077–1083
45. Rojas J, Salinas L, Garnica I (2003) Influence of the kinematic viscosity of oil contaminants in the compaction and hydraulic conductivity in certain type of soils. In: Groundwater engineering: recent advances: proceedings of the international symposium on groundwater problems related to geoenvironment, Okayama, Japan, vol 373. Taylor & Francis
46. Chew S, Lee C (2006) Simple shear behaviour of palm biodiesel contaminated soil. J Eng Appl Sci 5:12
47. Talukdar DK, Saikia BD (2013) Effect of crude oil on some consolidation properties of clayey soil. Int J Emerg Technol Adv Eng 3(2):117–120
48. Salimnezhad A, Soltani-Jigheh H, Soorki AA (2021) Effects of oil contamination and bioremediation on geotechnical properties of highly plastic clayey soil. J Rock Mech Geotech Eng 13(3):653–670

49. Najar IA, Ahmadi R, Amuda AG, Mourad R, Bendary NE, Ismail I, Bakar NA, Tang S (2025) Advancing soil-structure interaction (SSI): a comprehensive review of current practices, challenges, and future directions. J Infrastruct Preserv Resil 6(1):5
50. Walls WD (2010) Petroleum refining industry in China. Energy Policy 38(5):2110–2115
51. Ratnaweera P, Meegoda JN (2006) Shear strength and stress-strain behavior of contaminated soils. Geotech Testing J 29(2):133–140
52. NOSDRA (2021) Crude oil spill. Accessed on August, 2023 https://nosdra.oilspillmonitor.ng/
53. SPDC (2021) Crude oil spills. Accessed on June 2021 https://www.shell.com.ng/sustainability/environment/oilspills/Aug-2023.html

# Biobased Nanomaterials—A Green Remedy for Oil Spill Treatment

**Phaik Ching Ang, Veeradasan Perumal, Mohamad Nasir Mohamad Ibrahim, Sreeramanan Subramaniam, Arumugam Thangamani, Senthilkumar Obuliraj, and Pandian Bothi Raja**

**Abstract** Oil spills pose a substantial menace to maritime environments and ecosystems. Hence, the extraction of oil from water resources is a current and significant matter of concern. Among various spill oil removal methods, using sorbents, membranes, and aerogels to absorb and recover spilled oils is the prospective method. These products are cost-effective and able to reduce secondary pollution to the water resources compared to surface-active substances like dispersants, which are more toxic than oil. Currently, there is a growing interest in sorbents that are derived from biomass sources, because they are low-cost, abundant, biodegradable, and sustainable. Herein, several biomaterials on micro-to nanosized scale (cellulose-based, chitin/chitosan-based, and biochar-based) are discussed in this chapter. Various modification methods for biomass nanomaterials to enhance the spilled oil removal ability are also mentioned. This chapter encompasses the research undertaken by multiple scientists in the domain of eliminating and retrieving spilled oils by the utilisation of diverse biobased nanomaterials.

P. C. Ang · M. N. M. Ibrahim · P. B. Raja (✉)
School of Chemical Sciences, Universiti Sains Malaysia, Penang, Malaysia
e-mail: bothiraja@usm.my

V. Perumal
Centre of Innovative Nanostructures and Nanodevices (COINN), Universiti Teknologi PETRONAS (UTP), Seri Iskandar, Perak, Malaysia

Mechanical Engineering Department, Universiti Teknologi PETRONAS (UTP), Seri Iskandar, Perak, Malaysia

S. Subramaniam
School of Biological Sciences, Universiti Sains Malaysia (USM), Georgetown, Penang, Malaysia

A. Thangamani
Department of Chemistry, Karpagam Academy of Higher Education, Coimbatore, Tamil Nadu, India

S. Obuliraj
Department of Chemistry, Ramco Institute of Technology, Rajapalayam, Tamil Nadu, India

S. Subramaniam
Centre for Chemical Biology, Universiti Sains Malaysia, Penang, Malaysia

© The Author(s), under exclusive license to Springer Nature Singapore Pte Ltd. 2025 
S. A. Bhawani et al. (eds.), *Biobased Materials and Their Composites for Oil Spill Treatment*, Composites Science and Technology,
https://doi.org/10.1007/978-981-96-6049-0_4

# 1 Introduction

Global water contamination has resulted from fuel spills occurring throughout the whole lifecycle of petrol, including its manufacture, storage, and transportation. Additionally, leaks from underground storage tanks, refineries, oil production processes, fuel distribution terminals, and improper disposal have all contributed to this issue. Oil spills that enter the environment, especially the marine ecosystem, affect the immunotoxicity of marine life and seabirds [1]. When the oil reaches the shore, it also harms other creatures and terrestrial animals [2]. Hence, based on different cases, oil spills can be categorized into onshore and offshore oil spills. Oil spills can have long-term negative effects on aquatic life and the clean-up process is time consuming. Hence, cleaning up water resources polluted by oil spills (routine shipping, industrial effluents, or dumping), is a major challenge.

Conventional methods for treating oil spills include in-situ burning, physical/mechanical treatments (stationary/mobile floating devices, sorbents), chemical treatments (heat and UV, coagulation–flocculation, chemical disinfection), and biological treatments (biodegradation of oils by microbes) [3]. The pros and cons of each oil spill treatment methods are listed in Table 1. Most oil spill treatment methods mainly involve gravity separation, and chemical treatment have the problems of high operation cost and secondary pollution [4]. Hence, research on potential and environmentally benign bio-based products for treating oil spills in the form of sorbents, separators, dispersants, surfactants, and gelators has become a recent trend.

Bio-based nanomaterials are introduced to the treatment of oil spills. Nanomaterials are materials with size in the nanometre range, typically between 1 and 100 nm, while bio-based nanomaterials are those made from natural raw materials. In

**Table 1** Comparison of various oil clean-up methods [5–7]

| Classification | Example technology | Advantages | Limitations |
|---|---|---|---|
| Physical method | Booming and skimming<br>Use of sorbents<br>Mechanical removal | • Effective to remove quantities of oil<br>• Efficient<br>• Economically viable | • Time-consuming<br>• Safety concern<br>• Dependent upon condition at sca |
| Chemical method | Dispersants<br>Demulsifies<br>Solidifiers<br>Herders | • Simple operation<br>• Cheaper than physical method<br>• Suitable to treat large area | • Possible secondary pollution<br>• Labour intensive<br>• Environmental concern |
| Biological method | Bioremediation<br>Bio-stimulation | • Environmentally friendly<br>• Low operation cost | • Ineffective in large amount oil spill area<br>• Time consuming |
| Thermal method | In-situ burning | • Effective to remove quantities of oil<br>• Simple operation | • Possible secondary pollution<br>• Environmental concern |

recent years, numerous researchers and scientists have become interested in nanomaterials, and the field's research is advancing quickly, while bio-based nanomaterials and related nanocomposites are also increasingly gaining popularity because of their environmentally benign and renewable quality, low carbon footprints, and abundance of resources. Gelatin, albumin, alginates, chitin/chitosan, starch, and nanocellulose are a few examples of biobased nanomaterials [7]. In this chapter, the research and development of bio-based nanomaterials on oil spill treatment will be studied.

## 2 Cellulose Based Nanomaterials

The abundance of bio-based polymer (cellulose), nanomaterials that came from agricultural waste and forests has gained attention [7]. It possesses a variety of intriguing benefits, including biodegradability, biocompatibility, and thermal resistance, to emerge as a highly promising renewable polymer material [8]. Cellulose is a linear polysaccharide made up of non-water soluble β-1–4-linked D-glucopyranose repeating units [9, 10]. Other than the use of unmodified cellulose materials, cellulose can be converted into remodified cellulose materials, which are extensively used in a variety of industries.

In plants, around 36 individual cellulose molecules are joined together to create bigger structures known as elementary fibrils or microfibrils [9], then these latter structures are ultimately organised into familiar cellulose fibres. There are nine different types of cellulose nanoparticles (CNs) can be considered, which are distinguished from one another based on cellulose source materials and the techniques used to extract the particles. They are wood fibre (WF) and plant fibre (PF), microfibrillated cellulose (MFC), nano-fibrillated cellulose (NFC), microcrystalline cellulose (MCC), nanocrystalline cellulose (NCC), tunicate cellulose nanocrystals (t-CNCs), algae cellulose nanocrystals (AC), and bacterial cellulose particles (BC) [11]. Each type of particle is unique, with distinguishing features such as aspect ratio, size, crystallinity, morphology, and crystal structure. Among various types of cellulose, nanocellulose-based products (nano-fibrillated cellulose (NFC), bacterial nanocellulose (BNC), and nanocrystalline cellulose (NCC)) have shown high adsorption capacity due to their nanometre scale lateral dimensions, ultra-low density, and porous structure (Fig. 1) [7, 12].

To date, several natural organic materials, including kapok fibres, cotton, wood chips, sugarcane bagasse, and rice straws, have been suggested as potential and environmentally benign alternatives for oil recovery and clean-up sorbents due to their biodegradability properties [12]. One of the developments is the use of materials based on cellulose as natural oil sorbents. However, unmodified raw celluloses are amphiphilic and have a propensity to absorb water and oil unselectively [12]. Thus, a significant step in the production of oil sorbents is the transformation of amphiphilic cellulose into highly oleophilic/hydrophobic materials (Fig. 2) by altering their surface and morphology to boost their selectivity to oil is a crucial step to produce oil sorbents [14]. Furthermore, even though cellulose is biodegradable

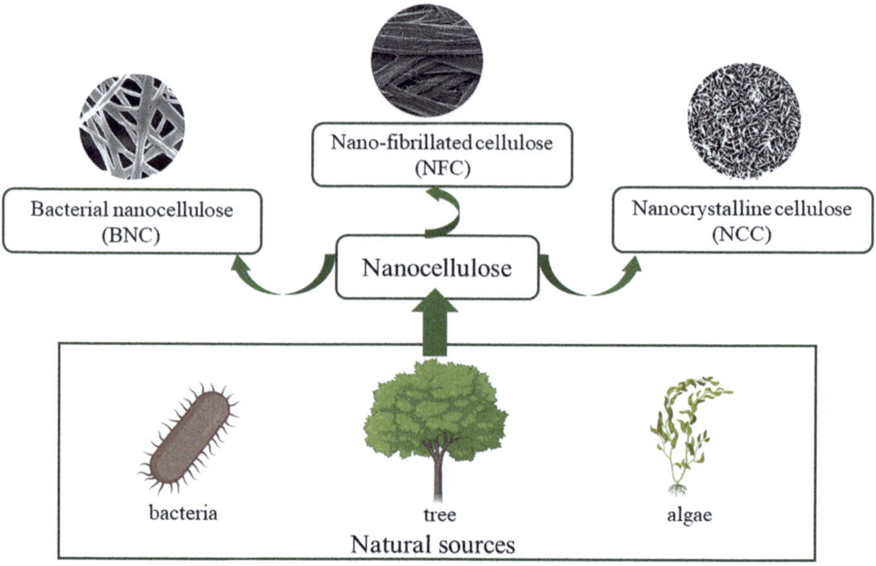


**Fig. 1** Nanocellulose categorized into NFC, NCC, BNC. Adapted from [13]

substance, reusability is still important to obtain the best sorbents instead of choosing single-use sorbents. There are two considerations that need to be taken into account for reusability, e.g., (i) the potential for cellulose material pores to deform under mechanical pressure, (ii) the materials' stability in the presence of high temperatures and solvents during chemical oil recovery [10].

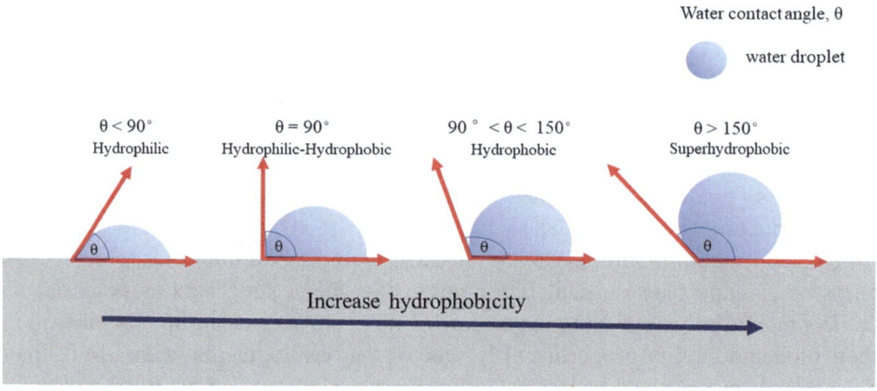


**Fig. 2** Wettability expression based on water contact angle on the surface of sorbent, θ

## 2.1 Cotton

Most studies of cotton-based sorbents were found for oil spill treatment. In a previous study by Wang et al., they reported the immobilization of silica nanoparticles on the superhydrophobic cotton fibre for oil sorption can improve the hydrophobicity of the fibre by rough structure [15]. The improvement was proved by the water contact angle (WCA) enhanced from below 90° to 153° after the coating with silica nanoparticles. The resulting fibre can absorb various oil, e.g., n-hexane, chloroform, toluene, paraffin oil, and crude oil, with more than 80 cycles without obvious loss of oil sorption capability. Another example from Shin et al. who substituted the free surface hydroxyl groups on untreated cotton with alkyl groups (specifically C18 fatty acid) effectively increased the fiber's resistance to water and affinity for oil [16]. The initial WCA of the raw cotton fibre was 110.70° and it increased to 138.10° after the modification. The enhancement can be further confirmed by approximately 1.5 times absorption capacity improvement compared to raw cotton fibre.

Instead of modification of cellulose surface, cellulose combined with aerogel is also one of the interesting topics for oil sorbents. Aerogels are light-weight solid and solvent-free materials that exhibit a high porosity (typically 95–99%) within their network structure [17]. Cellulose aerogels are a new class of high-efficiency oil sorbents that combine the unique characteristics of aerogels with biocompatible cellulose fibres as a potential alternative for environmental approaches (Fig. 3) [18]. Several examples of cotton cellulose-based aerogels derived from cotton and their performance are shown in Table 2.

Providing magnetic properties in cellulose aerogels by using magnetic materials based on iron oxide nanoparticles can also enhance the performance of sorbents [12].

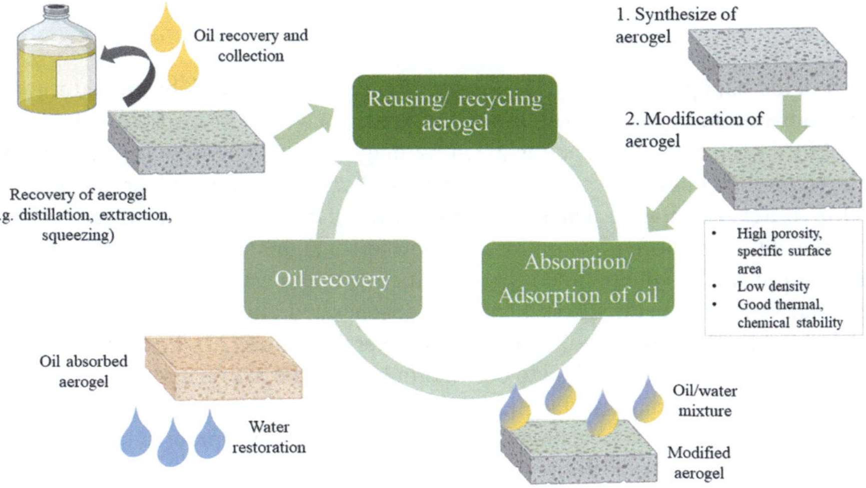


**Fig. 3** Natural bio-based aerogel in oil spill treatment

**Table 2** Cotton-based sorbents used for the removal and recovery of oil

| Sorbent | Synthesis | Types of oil | Water contact angle | Reusability cycles | References |
|---|---|---|---|---|---|
| Macro porous cellulose aerogel from raw cotton fibre | • One-step dissolution of raw cotton fibre<br>• Modifying the surface of aerogel with methyltrimethoxysilane (MTMS) using chemical vapor deposition process | Chloroform, n-hexane, toluene, gasoline, diesel, kerosene, DMF | >154° | ≈80 | [15] |
| Cotton-cellulose aerogel | • Mixing-blending cotton pad, cellulose fibres with polyamide-epichlorohydrin<br>• Aerogels are functionalized with MTMS | Dichloromethane, ethanol, acetone, motor oil | >130° | ≈5 | [19] |
| MTCS-modified cotton cellulose aerogel | • Chemical cross-linking of cotton cellulose solution<br>• Modification with methyltrichlorosilane (MTCS) by thermal chemical vapor deposition process | Chloroform, n-hexane, cyclohexane, pump oil, petroleum ether | >141° | ≈10 | [20] |
| Cotton-cellulose aerogel | • Dissolution of cotton cellulose and freeze drying<br>• Modification with trimethylchlorosilane (TMCS) by cold plasma method | Ethanol, diesel, gasoline, pump oil, peanut oil | >150° | ≈15 | [21] |

Iron oxide nanoparticles ($Fe_3O_4$), known as magnetite, are widely studied for their high adsorption capacity of oil, low cytotoxicity, magnetic properties, and biocompatibility [22]. Sorbents functionalized by magnetite also possess several benefits, e.g. (i) provide unsinkable properties for the sorbents to float on the surface of water/oil mixtures, (ii) able to achieve rapid removal and separation of aerogel under a permanent magnet after absorption, (iii) the ferromagnetic properties of magnetite able to transport to the polluted region and collect oil easily after adsorption by an external magnetic field, and (iv) could be regenerated after easy treatment include ultrasonic washing in ethanol or hexane for several minutes [22]. Peng et al. reported a simple co-deposition method to coat a thin layer of $Fe_3O_4$ on a medical absorbent cotton cellulose sponge surface [23]. $Fe_3O_4$ in this study is used to increase the roughness of the cellulose sponge surface. It shows high separation efficiency (more than 95%) and good reusability even after 5 cycles.

## 2.2 Recycle Cellulose Fibre

Recycled cellulose fibre, is also known as secondary cellulose fibres, such as paper [24]. The amount of paper being used has dramatically increased, resulting in large paper wastes that make up about 25–40% of garbage. Nguyen et al. prepared ultra-light, and highly porous aerogels from paper waste cellulose fibres [25]. The cellulose aerogel was functionalized with MTMS to enhance its hydrophobicity and oleophilicity and it resulted in WCA of approximately 145°. It exhibits high adsorption capacities of more than 18.4 g $g^{-1}$ of several crude oils where oil with lower viscosity can penetrate through the porous aerogel easier. Liu et al. created a cellulose-based oil adsorbent using wastepaper PS-16, specifically mixed kraft cuttings. The object was treated with alkyl ketene dimer and polyethylene wax using a spraying process in order to increase its hydrophobicity, as indicated by a water contact angle (WCA) of 125.6°. The adsorbent had the oil recovery up to 92.8% with a capacity of 16–28 times its weight.

Some researchers introduced composite sorbent to enhance the performance of recycled cellulose in oil clean-up methods. For example, Parathap et al. developed a composite sorbent by impregnating 1,2:5,6-di-*O*-cyclohexylidene-mannitol (phase selective organogelator) into cellulose pulp (solid carrier) [26]. The initial purpose of the organogelator was to temporarily render the sorbents hydrophobic by forming hydrogen bonds with hydroxyl groups on the cellulose pulp surface, thereby revealing the hydrophobic area (WCA: 110°). Furthermore, this sorbent could quickly absorb the oil through capillary action and release the organogelator into the absorbed oil for its self-assembled gelation. It could prevent the oil from spreading and coagulated it into an easily collectible rigid floating mass. As a result, it could absorb various crude oils with different viscosities by 16 times its weight.

## 2.3 Nano-fibrillated Cellulose (NFC)

Fibrils are present in the outer layers of the helically organised wall structure of chemical pulp fibres. This wall structure is characterised by an unique pattern created by wrinkles [27]. NFCs are semicrystalline spaghetti-like nanoparticles with a thickness of 2–60 nm and a length of a few millimetres with a small amount of hemicellulose (Fig. 4) [28]. NFCs are derived from delamination of wood pulp by mechanical pressure [6]. Herein, there are a few examples of studies on fibril sorbents from various kinds of native sources.

Chhajed et al. fabricated NFC/PVA-SAC-based aerogel by physically cross-linking between NFC from mango wood scraps and polyvinyl alcohol (PVA). The stearic acid (SAC) coating was used to modify this aerogel, which has a water contact angle (WCA) greater than 159°, making it superhydrophobic. This aerogel able to reuse at least 15 cycles while maintaining high absorption efficiency. The aerogel

**Fig. 4** Illustration of NFC structure. Adapted from [29]

could be reused for several cycles as the tough nature of PVA exhibited elastic behaviour, for the aerogel can easily apply simple mechanical squeezing.

Titanium dioxide ($TiO_2$) is introduced to achieve hydrophobicity and oleophilicity in several research. Kettunen et al. reported a light-weight nanocellulose aerogels with hydrophobic surfaces by the functionalization of $TiO_2$ coating on them [17]. The nanocellulose aerogels are formed from long and entangled native cellulose nanofibrils due to strong mutual hydrogen bonds. However, the aerogels are superabsorbent materials and are unable to be applied in oil spill treatment. Thus, $TiO_2$ is also used as stabilizer and provides photo-switchable wetting process for the aerogels between water-superabsorbent and water-repellent states. Korhonen et al. used nanofibrils from hardwood kraft pulp to produce a hydrophobic nanocellulose aerogel with a small density of 20–30 mg $cm^{-3}$ and porosity of more than 98% [30]. This aerogel is also functionalized by $TiO_2$ using atomic layer deposition to have hydrophobicity properties (WCA > 90°). The aerogel could be simply dried both in a vacuum oven at 60 °C or in ambient conditions, it worked well equally for the recovery process.

## 2.4 *Nanocrystalline Cellulose (NCC)*

NCCs, also known as nano-whiskers, are highly crystalline rigid rod-like nanoparticles with a thickness of 5–70 nm and a length of a few hundred nanometres [6, 28]. NCCs are generated through the elimination of non-crystalline sections from various cellulose sources via acid hydrolysis (Fig. 5) [6, 31]. Typically, a strong acid, i.e., hydrochloric acid (HCl) or sulfuric acid ($H_2SO_4$), is employed to destroy the amorphous domains of cellulose, leading to the formation of NCCs [12]. The attractive features of NCC are the different morphologies and dimensions, outstanding mechanical properties, low density, high surface reactivity, and self-assembling properties [32].

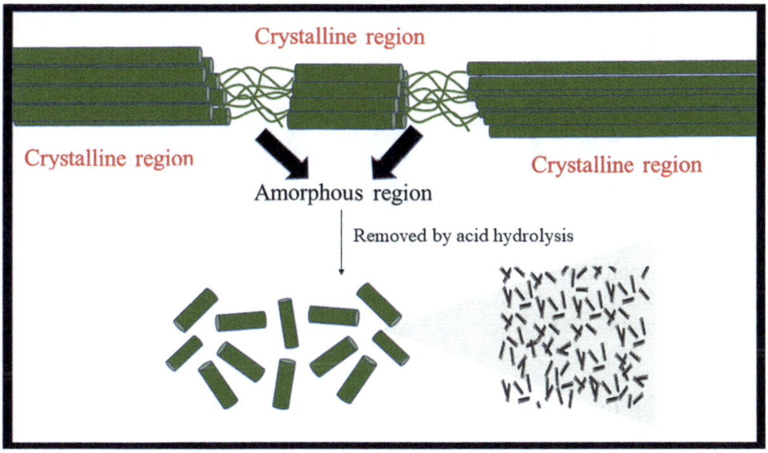


**Fig. 5** Forming NCC structure after removing amorphous region by acid hydrolysis

Huang et al. prepared superhydrophobic wood substrates by two spraying processes, i.e., (1) commercial spray paint was used as an adhesive on substrates, and (2) the modified NCC was sprayed onto the adhesive [33]. This sorbent gave a WCA of 163° with high water-repellence. Besides, it also demonstrated good reusability, high separation efficiency, and excellent corrosion resistance properties, because the aggregation of NCC on the substrate generated irregular particles and led to the formation of rough surface morphology. The oil separation of this sorbent was above 97% and was able to repeat approximately 40 cycles.

In the research of Akhlamadi et al., they created a superhydrophobic, eco-friendly, and recyclable aerogel from the aqueous solution of waste tissue paper-based NCCs and polyvinyl alcohol (PVA) in the presence of hydrolysed tetraethyl orthosilicate (TEOS) by freeze-drying [34]. The prepared aerogel was ultralight (0.017 g $cm^{-3}$) with high porosity (98.42%) and a WCA of approximately 154.93°. These results further support the obtained aerogel with high sorption capacity (168 g $g^{-1}$) for 14 types of oils and organic solvents. Moreover, the aerogel could maintain 92% of its sorption capacity after 20 cycles, and 89% of shape recovery after 50 cycles, which indicated the high potential of reusability.

## *2.5 Bacterial Nanocellulose (BNC)*

BNCs are produced using a bottom-up strategy via constructing NFCs from simple sugars using microbial fermentation (Fig. 6) [28]. BNCs assembled in ribbon-shaped fibrils with a width of less than 25–100 nm and thickness of 20–100 nm [6]. The unique properties of BNCs enable them to be used in oil spill treatment methods, including high purity, biocompatibility, elasticity, absorbency, and high

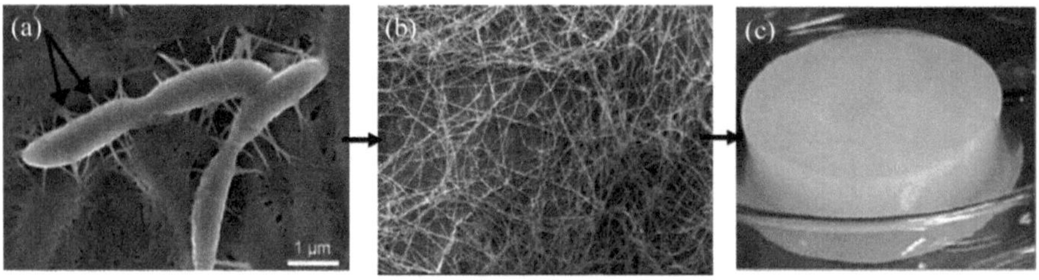


**Fig. 6** Bottom-up approach to produce BNC. **a** Forming nanocellulose fibre by *Gluconacetobacter bacteria*, **b** SEM micrograph of BNC with a 3D network, **c** BNC hydrogel. Reproduced from [37] (Available via license: CC BY 4.0)

tensile strength in the wet state [35]. Sai et al. produce the 3D web-like structure of bacterial cellulose aerogels (BCA) by modifying the bacterial cellulose hydrogel i.e. *nata-de-coco* to [36]. In order to obtain hydrophobic properties, the surface was treated with a solution of trimethylchlorosilane (TMCS) in a mixture of triethylamine and dichloromethane, and then subjected to freeze-drying. The BCA has a WCA of more than 146.5° and can collect various oils and organic solvents, i.e., chloroform, toluene, diesel, dichloromethane, paraffin oil, and plant oil, up to 185 g $g^{-1}$. This BCA can be reused after compression and rinsing with tert-butanol.

In the research of Ieamviteevanich et al., they functionalized bacterial cellulose aerogels by co-precipitation of $Fe_3O_4$ nanoparticles before pyrolysis to produce magnetic carbon nanofiber aerogel that decorated with Fe/$Fe_3O_4$ core–shell structure [38]. The highly dispersed core–shell structure nanoparticles and no obvious aggregation were observed. This aerogel provides a relatively large absorption capacity (37 g $g^{-1}$ to 87 g $g^{-1}$) for a wide range of oils. However, they discovered that despite increasing the nanoparticle concentration, agglomeration of cluster particles continued to occur, significantly decreasing the aerogel's hydrophobicity. Forming protecting shells, e.g. surface coatings with polymers, surfactants, and organic or inorganic compounds on the surface of nanoparticles, is considered a good way to isolate and functionalize the core of nanoparticles against [39] the environment [22].

Although BNC-based sorbents give good performance in oil clean-up applications, they are relatively expensive due to their low productivity and the use of expensive culture media. Hence, BNC composite with cheaper plant cellulose is also a method to reduce the production cost on an industrial scale. Wang et al. successfully integrated BC with a more affordable type of plant cellulose called needle-leaf bleached kraft pulp (NBKP) to enhance the porosity of the composite membrane using the dip coating technique, hence reducing the overall cost [39]. BC/NBKP composite was modified by copper and stearic acid to promote the oil super wettability and improve separation efficiency. It displayed a petal-like micro-structure with a WCA of 162.3° and separation efficiency of more than 95% for various oils. It could be reused for at least 10 cycles after washing with ethanol.

## 2.6 Other Cellulose Biomass

There are many agricultures cellulose by-products worldwide, such as sugarcane bagasse [40], banana peels [41], corn stalks [42], kapok fibres [43], barley straws [44], oil palm fibres [45], pomelo peels [46], rice husks [47], sawdust [48], wheat straws [49], etc. are produced daily. These natural plant wastes indicate there are still many possibilities for researchers to the path of develop excellent oil clean-up methods. The benefits of using these wastes are their ecological safeness, high possibility modification methods, and relatively cost-effective. Hence, understanding their physical properties, chemical properties, structure, and components becomes very important before applying them to oil spill treatment applications.

Based on Ali et al., the water/oil sorption effectiveness of kapok fibres, rice husks, and sugarcane bagasse was compared [50]. Based on the EDX result, kapok fibres and sugarcane bagasse consist of a carbon percentage greater than 50%, which indicates most components are organic substances (cellulose, lignin, hemicelluloses). Rice husks consist of above 20% amorphous silica ($SiO_2$) due to the presence of opaline silica. The density was increased from kapok fibres (0.7358 g $cm^{-3}$), sugarcane bagasse (0.9866 g $cm^{-3}$), and rice husks (1.1575 g $cm^{-3}$) (Fig. 7). Rice husks show relatively weak sorption capacity in oil (<10 g $g^{-1}$) and water (<5 g $g^{-1}$). Sugarcane bagasse exhibits comparable high water sorption capacity (>5 g $g^{-1}$) and oil sorption capacity (>10 g $g^{-1}$). Kapok fibres show the best oil sorption capacity (>15 g $g^{-1}$) and weakest water sorption capacity (<1 g $g^{-1}$) among these three cellulose biomasses. The result indicates untreated kapok fibres provide the best hydrophobicity properties; untreated sugarcane bagasse was moderately recommended before applying in oil clean-up sorbents, whereas rice husks were not recommended without any modification.

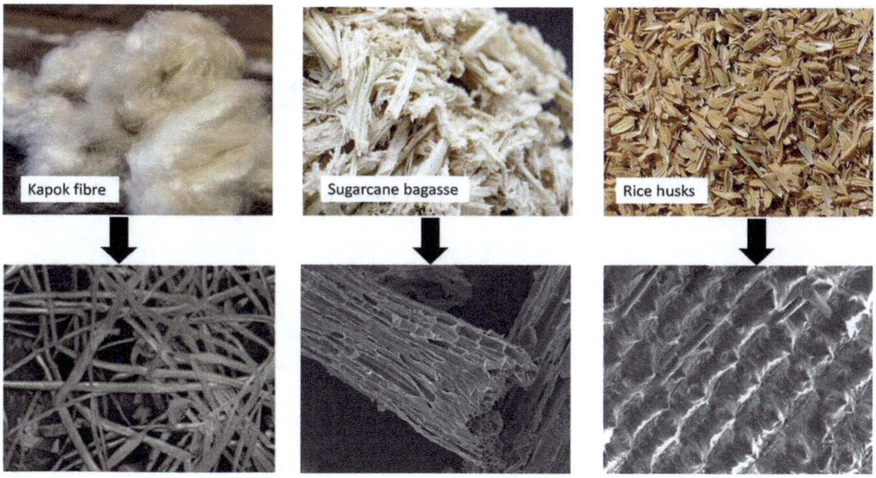


**Fig. 7** Kapok fibres, sugar bagasse, rice husks, and their SEM micrograph

According to the report by Kenes et al., the nonporous structure of untreated rice husks was the reason for their low sorption capacity [47]. They modified the rice husk by heating it to 700 °C in the muffle furnace under $CO_2$ flow. The result showed when the rice husk thermally increased up to 700 °C, the specific surface area increased with a decrease in density. Hence, the heated rice husk ($TRH_{700}$) sorbent obtained could provide a higher sorption capacity (15 g $g^{-1}$) for petroleum compared to virgin rice husk (2 g $g^{-1}$).

## 3 Chitin/Chitosan Based Nanomaterials

Chitin is a natural polysaccharide consisting of repeated *N* acetyl-D-glucosamine units linked by β-(1,4) glycosidic bonds [51, 52]. It is the second most abundant polymer after cellulose, mainly found in crustaceans, insects, and microorganisms (Fig. 8). Glucans and proteins, which frequently make up of natural chitin with more than 50% of the mass in chitin-containing tissue [53]. Typically, chitin exists in three crystalline polymorphic forms (α-chitin with parallel chains, β-chitin with anti-parallel chains, and γ-chitin with a mixture of parallel and anti-parallel chains) [51, 53]. The most popular chitin is α-chitin, which contains ordered crystalline microfibrils that serve as structural elements in the exoskeleton of arthropods (shrimp, crab) as well as in the cell walls of fungi and yeast [54]. Chitin is low toxicity, biodegradable, biocompatible, and capable of defending against bacterial invasion by chitinases.

The most famous derivative of chitin, chitosan was produced after a chitin deacetylation process, which convert the acetyl groups (–$COCH_3$) into an amine group (–$NH_2$) (Fig. 9) [51]. The deacetylation methods could be enzymatic hydrolysis in the presence of an enzyme or chemical hydrolysis under alkaline conditions [56].

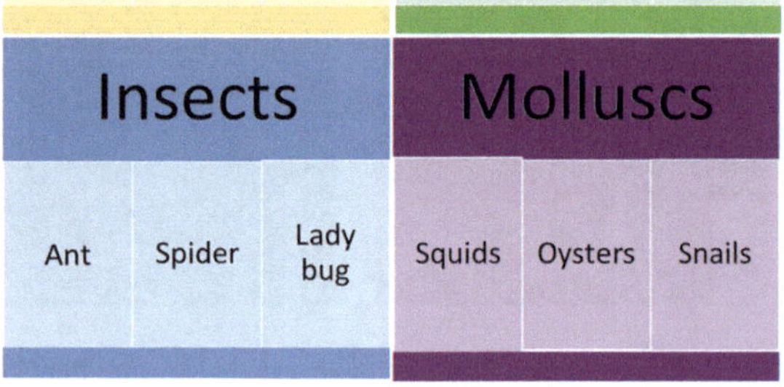


**Fig. 8** Natural sources of chitin. Adapted from [55] (Available via license: CC BY 4.0)

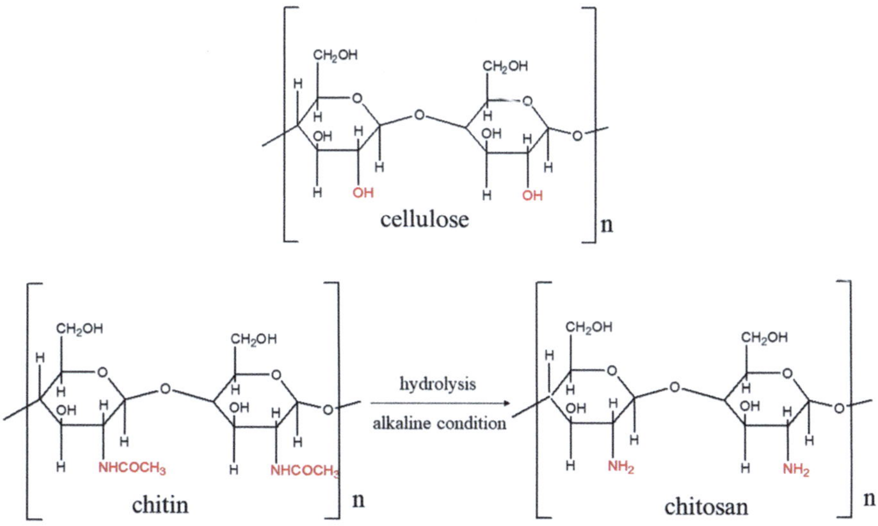


**Fig. 9** Comparison of chemical structure of bipolymers (cellulose, chitin, chitosan)

Chitosan is a copolymer of *N*-acetyl-D-glucosamine and D-glucosamine, whose total nitrogen content is more than chitin. The nature of chitosan (one amine group and two hydroxyl groups) makes it possess a higher chemical reactivity than chitin [57]. It was able to dissolve in an acidic medium that was lower than pH 6 [58]. There are some drawbacks of chitosan to produce on industrial scale, e.g., limited suppliers, low thermal stability, mass transfer resistance, low porosity, and surface areas [56, 58]. However, the presence of amino and hydroxyl groups in chitosan, increases the possibilities of chemical modifications it and further utilization in oil spill treatment applications, e.g. alkylation, acylation, oligomerization, carboxy alkylation, hydroxy alkylation, thiolation, graft copolymerization and enzymatic modifications [58].

As mentioned, chitin is recovered from natural sources, especially shrimp waste and crab shells. The basic method follows a four-step process, i.e., (i) extraction of chitin from raw resources, (ii) removal of calcium carbonate by demineralization using dilute HCl, (iii) deproteination using dilute sodium hydroxide aqueous (NaOH), (iv) decolorization [51]. Meanwhile, continuing the third process with the deacetylation process will produce chitosan [57]. Chitosan is produced when the deacetylation degree exceeds 0.65 [53]. The schematic process refers to Fig. 10. Furthermore, chitin and chitosan can be produced in nano-size via phase separation, electrospinning, template synthesis, and self-assembly, leading to a wide range of potential applications [51].

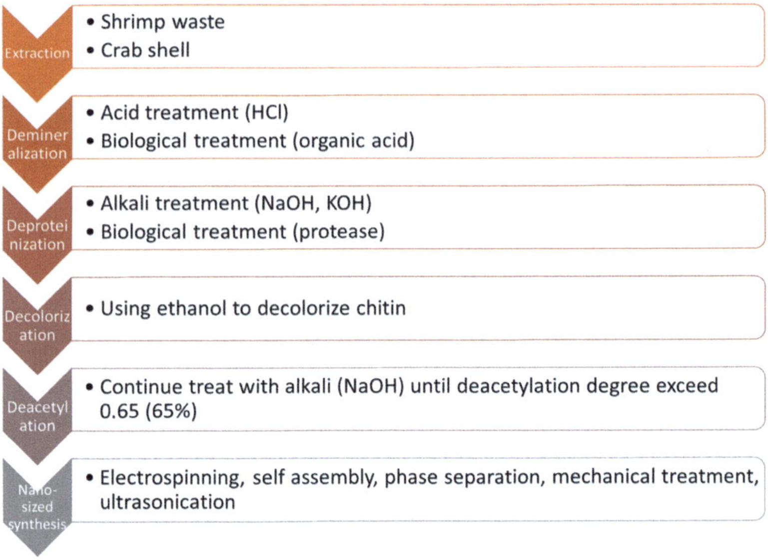


**Fig. 10** Overall process in the production of chitin and chitosan nanomaterials [51, 59]

## *3.1 Shrimp Waste*

Nowadays, shrimp waste has undergone a dramatic increase worldwide, especially in Asia, which plays a role in almost 80% of shrimp production [59]. About 40–48% of shrimp by weight (head and body carapace) was being discarded [59]. It became a serious environmental problem because valuable raw bio-sources (protein, minerals, chitin) in shrimp wastes are thrown away. Hence, there is a significant to recycle shrimp waste in various applications. For example, the relative abundance and high protein content of shrimp wastes ensure it becomes a promising resource for producing chelating, emulsifying, and adsorbent materials for oil clean-up methods.

In the study of Zhang et al., they hydrolysed the grounded shrimp waste enzymatically by alcalase enzyme to produce a green dispersant [60]. This dispersant could reduce the water surface tension until 33.23 dynes $cm^{-1}$ at a critical micelle concentration (CMC) of 19.9 g $L^{-1}$. Furthermore, the shrimp waste dispersant able to dissolve in water as a solvent, indicates the utilization of the product has benefits such as higher dispersant effectiveness, lower toxicity, cost-effectiveness, and higher stability under different conditions. As a result, it provides effective oil dispersion for several crude oils even in cold seawater. Piotrowska-Kirschling et al. synthesized the polyurethane-chitosan foams (PUR-Ch) by two-step method [61]. First, the PUR prepolymer was synthesized from polyol and diisocyanate in the presence of a catalyst. Then, the PUR prepolymer was foamed with water and modified with 1.5% weight of chitosan, which was obtained from the deacetylation of extracted chitin

**Table 3** Study of chitosan and chitosan coagulant mixtures for oil removal [62]

| Sorbate | % mixture | % oil removed |
|---|---|---|
| Chitosn | 100 | 20 |
| CMC | 100 | 80 |
| $Al_2(SO_4)_3$ | 100 | 60 |
| Chitosan/CMC | 90:10 | 99 |
| Chitosan/$Al_2(SO_4)_3$ | 80:20 | 85 |

from shrimp waste, simultaneously. The oil sorption capacity of this foam could reach above 450% after 1 day of sorption with at least three cycles.

Hosny et al. also introduced the synthesized chitosan from extracted shrimp shell chitin combined with different coagulants, i.e., carboxy methyl cellulose (CMC) and aluminium sulphate ($Al_2(SO_4)_3$) to treat with oily water [62]. The mechanism in this study is the coagulation of small particles into large aggregates by charge neutralization and destabilization of colloidal impurities, then the oil aggregates could be removed easily by filtration and sedimentation. Chitosan/CMC with ratio 9:1 could remove almost 99% of oils, whereas chitosan/$Al_2(SO_4)_3$ with a ratio 8:2 could remove 85% of oils. The result in Table 3 shows the ability of chitosan to enhance oil removal after adding small coagulant ratios, especially chitosan/CMC. Hence, the amount of coagulants and their cost can be reduced by replacing them with a large amount of chitosan.

## 3.2 *Crab Shell*

Crab shells are also one of the major by-products of the seafood industry. The shell is the exoskeleton of crabs with multifunctional, such as supporting their body, resisting desiccation, providing environmental protection, and resisting mechanical loads. According to Gadgey et al., the crab shell consists of three parts, which are the epicuticle (outermost layer), exocuticle (outer/middle), and endocuticle (inner) [63]. Epicuticle is a thin, waterproof layer; exocuticle and endocuticle (procuticle) are the main structural parts of the crab shell to resist mechanical loads. The long chain polysaccharide chitins of crab shells form fibrils with 3 nm in diameter and 300 nm in length. Additionally, there is also amorphous calcium carbonate ($CaCO_3$) that is deposited inside the chitin-protein matrix in the crab shell [63].

Chitosan from crab shell wastes was suggested by Roy et al. to be used as an eco-friendly superhydrophobic coating onto a polyester fabric or medical grade cotton [64]. The obtained WCAs were 158.6° and 161.4° respectively. Based on the FESEM images, the chitosan powder with an average size of 200 nm was dispersed on the surface of the sorbent. The rough surface could transform the hydrophilic become an excellent hydrophobic sorbent to separate a variety of oils from the oil/water mixtures. As a result, the chitosan coated sorbent could be used at least 80 cycles to remove various oils, e.g., hexane and heptane. In another study from Doshi et al., they synthesized water soluble N, O-carboxymethyl chitosan (N, O-ChS) by carboxymethylation process between crab shells chitosan and chloroacetic acid in a hydro-alcoholic medium at 50 °C [65]. The N, O-ChS could destabilize three types of crude oils into small droplets and adsorb them at alkalinity and salinity of sea water level. The dosage of N, O-ChS in the range of 5–25 mg $g^{-1}$ was enough to treat the oils with different viscosity.

# 4 Other Bio-Based Nanomaterials

Cellulose and chitin/chitosan are the mainly discussed sustainable and renewable biomass that utilized in oil spill treatment applications. However, there are still many other examples that have widespread attention among scientist, such as starch, biochar, alginates, gelatin etc., due to their biodegradable, renewable, and a low carbon footprint property.

## *4.1 Starch*

Starch is a major dietary and natural source of carbohydrate polymer, which is able to be extracted from rice, corn, maize, wheat, potato, barley, soya, etc. It contains amylose, i.e., linear molecules linked to two D-glucose units through α-1, 4 glycosidic bonds, and amylopectin, i.e., branched chain molecules with branches occurring at α-1, 6 glycosidic bonds [66]. There are three types of starches, including type A (outer double helical structure with 23–29 glucose units, mostly found in cereals), type B (structure consists of 30–44 glucose units with water interspread, mostly found in potatoes), and type C (a combination of type A and type B, mostly found in beans) [67].

Without the need for cross linkers, starch can be dissolved in hot water easily and formed into an integrated hydrogel network structure [68]. This property enables its recyclability when utilized in sorbents without using chemical solvents and the encapsulation of some functional components to prevent secondary pollution.

Wang et al. proposed the possibility of modifying the surface of starch-based sorbents to become hydrophobicity in several studies. One of the studies focused on modifying the porous, magnetism, and hydrophilicity surface of potato starch cryogel with magnetite and silica nanoparticles (HMS-$SiO_2$@MSC) [68]. The WCA of the cryogel was 154.4°, and the oil sorption capacity in various organic liquids (range: 2.7–7.8 g $g^{-1}$) indicated the super hydrophobicity surface was formed successfully. It exhibited excellent water-repellent, anti-fouling, and self-cleaning properties to remove oil underwater and magnetically collect the oil slick on the water surface repeatedly for ~50 cycles without a dramatic change of WCA. In another investigation conducted by Wang et al., a solution of methyltrichlorosilane (MTS) was applied to the potato starch cryogel using a simple immersion process. This resulted

**Table 4** Comparison of WCA by using different coatings [70]

| Coatings | Uncoated | PDMS | EHR-SNPs | NP-SNPs | EHR-SNPs@PDMS | NP-SNPs@PDMS |
|---|---|---|---|---|---|---|
| WCA (°) | 52.85 | 109.22 | 90.78 | 25.40 | 151.85 | 152.4 |

in the formation of a honeycomb coral-like structure and an enhancement in the hydrophobic properties of the cryogel surface [69]. The modification of MTS has been proved to be a facile strategy to provide super hydrophobicity properties on substrate with the low-surface-energy and hierarchical micro/nanostructures. The WCA of this cryogel was above 151°, and the oil sorption capacity in various organic liquids was in the range of 2.5–7.6 g $g^{-1}$). However, this method is unable to show good recyclability since the WCA dropped to ~130° after 50 cycles.

Instead of modifying starch-based sorbents, starch itself in nano size could be the coatings to improve the hydrophobicity of another type of substrate to absorb/adsorb oils. To date, most research on super hydrophobic coatings relies on the use of non-degradable inorganic components [70]. Therefore, the fabrication of super-hydrophobic coatings with bio-based materials has attracted significant research interest. Wang et al. fabricated the composite coatings by ultrasonically combining polydimethylsiloxane (PDMS), i.e., a biodegradable, non-toxic, and low surface energy silicone substance, with starch nanoparticles (SNPs) from maize starch. Recrystallization (EHR) and nanoprecipitation (NP) were used as a greener pathway to prepare the SNPs. The WCA improvement is shown in Table 4.

## 4.2 *Alginates*

Alginates are anionic polysaccharides copolymers made up of 1,4 linking β-D-mannuronic acid (M) and α-L-guluronic acid (G). These monomers are arranged in blocks of repeating G residues (GG blocks), blocks of repeating M residues (MM blocks), and blocks of mixed M and G residues (MG blocks) (Fig. 11) [71, 72]. The polymer chain backbone is abundantly dispersed with free –OH and –CO groups. Naturally, it can be found inside brown algae or exopolysaccharide of bacteria, e.g., *Pseudomonas aeruginaso*. Alginates are biocompatible, biodegradable, non-immunogenic, and non-toxic. Alginate particles in nanometres have various advantages over alginate particles in micrometres, e.g., stronger mechanical strength and a bigger specific surface area. However, the majority of alginate particles mentioned in the literature have a diameter in micrometres. Consequently, there is growing attention to synthesize alginate nanoparticles. Typically, alginate nanoparticles formation is based on three methods: (i) complexation, (ii) electrospinning, and (iii) emulsification [73, 74].

In order to remove oil from aqueous solutions, Eskhan et al. developed calcium alginate hydrogel that was treated with maleic anhydride via esterification [76]. The WCA was increased from 83.7° to 118.9°, and the adsorption capacity of oil was

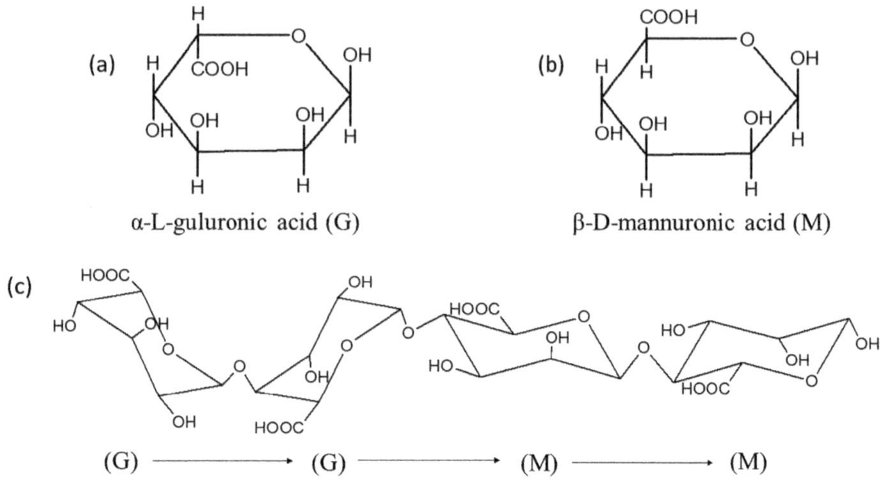


**Fig. 11** Chemical structures of **a**, **b** alginate monomers (G, M) and **c** polymer showing blocks with different repeating residues [75]

increased twice after the surface of native hydrogel was modified with maleic anhydride. Furthermore, it could be reused for at least 5 cycles with minimal losses by using solvent regeneration. Interestingly, alginate can also form hydrophilic substrates after modification to recover the water from oil inversely. Zhuang et al. fabricated a highly efficient alginate nanocomposite aerogel via freeze-drying method to remove various oils [77]. They incorporated titanium oxide/graphene oxide ($TiO_2$/RGO) nanocomposite into alginate with a ratio of 10:1:100 to form the aerogel (TRGA). In a result, a rough surface nanostructure was formed and displayed an oil contact angle reach to 161.2°. This showed the aerogel is hydrophilic/oleophobic with the ability to permeate water molecules and repel, and retain the oil droplets to achieve the separation. TRGA aerogel showed 99.96% of oil/water separation efficiency and maintained the effectiveness for more than 60 cycles due to anti-fouling properties.

## 4.3 *Biochar*

A variety of carbonaceous materials are available for cleaning up oil spills, e.g., carbon nanotubes, activated carbon, graphene-based compounds, and carbon aerogels [78]. Due to the availability of the raw feedstock, cost-effectiveness preparation, physicochemical characteristics, and universal oil clean up techniques, biochar has emerged as an interesting topic among those carbonaceous materials [79]. The product is a stable and porous material with a high carbon content. It is obtained from biomass using thermal treatment methods such as slow pyrolysis, hydrothermal

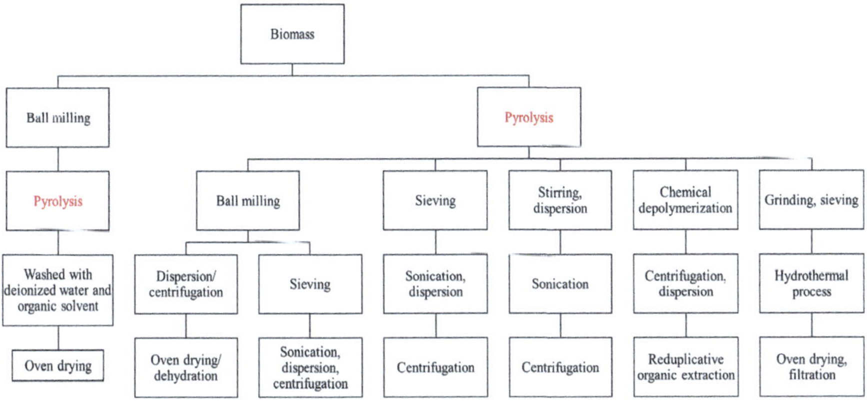


**Fig. 12** Approaches to obtain biochar and nano-biochar [82]

carbonisation, flash carbonisation, torrefaction, and gasification, which involve low amounts of oxygen. [78, 80]. Pyrolysis is the most popular technique to produce biochar (Fig. 12). Ash, trace amounts of nitrogen (N) and sulphur (S), carbon (C), hydrogen (H), and oxygen (O) are the components of biochar [81].

There are two types of biochar, including bulk biochar (particle sizes at 0.04–20 mm) and nano-biochar (particle sizes $< 100$ nm) [82]. According to Madhubashani et al., extreme pyrolysis temperatures over 400 °C to produce bulk biochar could reduce the oil removal efficiencies because of the damage of hole structure in raw materials, and further increased the temperature above 700 °C causing the loss of functional groups in the structure [78]. Hence, the smaller size of particles and higher specific surface area of nano-biochar could solve the drawbacks of using bulk biochar directly in applications (Fig. 13). The most common preparation of nano-biochar is using ball milling, which can be referred to Fig. 12.

Duan et al. fabricated a high-performance modified foam (BMBC@MF) by loading ball-milled wood chips biochar (BMBC) that was prepared at 300 °C and octadecylamine on melamine foam (MF). The purpose of BMBC in this study was to form nano/micron-scale folds to reduce the surface energy. This foam could absorb up to 43–155 times its weight due to its high porosity and hydrophobicity. The WCA was 155° when the loading BMBC suspension was 3 mg $mL^{-1}$ onto the MF skeleton. Further increasing the loading decreased the roughness and hydrophobicity of the surface. It could be reused for at least 11 cycles after the easy squeezing recovery method.

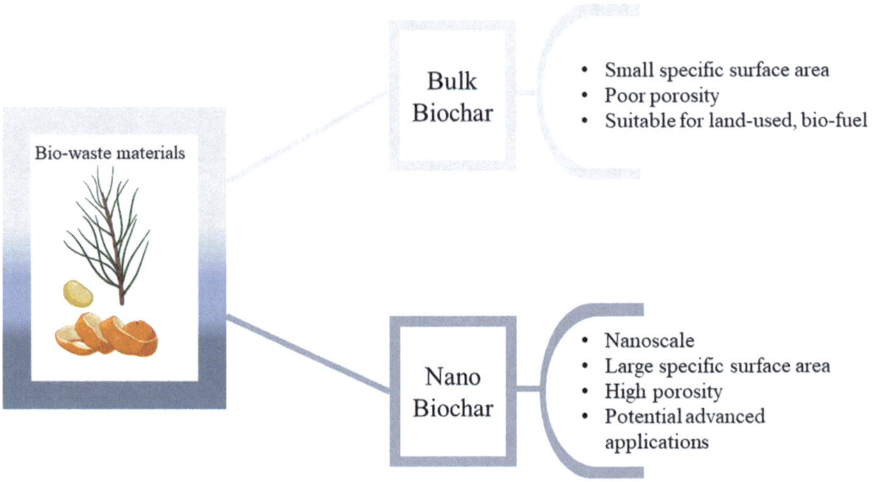


**Fig. 13** Illustration of properties comparison between the bulk biochar and nano-biochar

## 5 Summary and Conclusion

A large variety of nanocomposites or nanomaterials are known to date for oil spill treatment. However, it is still difficult to choose the most appropriate and eco-friendly materials or methods for separating oil from water completely. There are several factors can influence the final choice, e.g., availability or simplicity of its fabrication, the cost of production of a nanomaterial, reusability or reproducibility of nanomaterial time-by-time, stability in natural conditions (temperature, humidity, salinity of water), and type of application [3]. Due to the small particle size and strong reactivity of nanomaterials, they also being concerned about the toxicity that may harm the microbial communities inside the human body through ingestion, inhalation, or absorption through the skin [83]. Consequently, several biocompatible, less harmful, natural nanomaterials for treating oil spill contamination are presented in this chapter. The discussed examples demonstrate that these bio-based nanomaterials will degrade after their intended usage without worsening the state of the environment.

**Acknowledgements** The authors express their gratitude to the Ministry of Higher Education Malaysia (MOHE) for providing financial support through the Fundamental Research Grant Scheme (FRGS) with the code: FRGS/1/2021/STG05/USM/02/4.

## References

1. Barron MG (2012) Ecological impacts of the deepwater horizon oil spill: implications for immunotoxicity. Toxicol Pathol 40(2)

2. Brody TM, Bianca P Di, Krysa J (2012) Analysis of inland crude oil spill threats, vulnerabilities, and emergency response in the Midwest United States. Risk Anal 32(10)
3. Kharisov BI, Rasika Dias HV, Kharissova OV, Méndez YP (2017) Nanomaterial-based methods for cleaning contaminated water in oil spill sites. In: Nanotechnology for energy sustainability. Wiley-VCH Verlag GmbH & Co. KGaA, pp 1139–1160
4. Liu X, Ruan W, Wang W, Zhang X, Liu Y, Liu J (2021) The perspective and challenge of nanomaterials in oil and gas wastewater treatment. Molecules 26. MDPI AG
5. Behnood R, Anvaripour B, Jaafarzadeh N, Farasati M (2016) Oil spill sorption using raw and acetylated sugarcane bagasse. J Cent South Univ 23(7)
6. Liu H, Geng B, Chen Y, Wang H (2017) Review on the aerogel-type oil sorbents derived from nanocellulose. ACS Sustain Chem Eng 5(1)
7. Bagal-Kestwal DR, Kestwal RM, Chiang BH (2016) Bio-based nanomaterials and their bionanocomposites. Nanomater Nanocompos
8. Zhang H, Liu T, Zhu Y, Hong L, Li T, Wang X et al (2020) Lipases immobilized on the modified polyporous magnetic cellulose support as an efficient and recyclable catalyst for biodiesel production from Yellow horn seed oil. Renew Energy 145
9. Sun S, Sun S, Cao X, Sun R (2016) The role of pretreatment in improving the enzymatic hydrolysis of lignocellulosic materials. Bioresour Technol 199
10. Zamparas M, Tzivras D, Dracopoulos V, Ioannides T (2020) Application of sorbents for oil spill cleanup focusing on natural-based modified materials: a review. Molecules 25
11. Moon RJ, Martini A, Nairn J, Simonsen J, Youngblood J (2011) Cellulose nanomaterials review: structure, properties and nanocomposites. Chem Soc Rev 40(7)
12. Hammouda S, Chen Z, An C, Lee K (2021) Recent advances in developing cellulosic sorbent materials for oil spill cleanup: a state-of-the-art review. J Cleaner Prod 311
13. Klemm D, Cranston ED, Fischer D, Gama M, Kedzior SA, Kralisch D et al (2018) Nanocellulose as a natural source for groundbreaking applications in materials science: today's state. Mater Today 21
14. Łukawski D, Lisiecki F, Dudkowiak A (2018) Coating cellulosic materials with graphene for selective absorption of oils and organic solvents from water. Fibers Polym 19(3):524–530
15. Wang J, Geng G, Wang A, Liu X, Du J, Zou Z et al (2015) Double biomimetic fabrication of robustly superhydrophobic cotton fiber and its application in oil spill cleanup. Ind Crops Prod 77
16. Shin Y, Han KS, Arey BW, Bonheyo GT, Bonheyo GT (2020) Cotton fiber-based sorbents for treating crude oil spills. ACS Omega 5(23)
17. Kettunen M, Silvennoinen RJ, Houbenov N, Nykänen A, Ruokolainen J, Sainio J et al (2011) Photoswitchable superabsorbency based on nanocellulose aerogels. Adv Funct Mater 21(3)
18. Gu H, Zhou X, Lyu S, Pan D, Dong M, Wu S et al (2020) Magnetic nanocellulose-magnetite aerogel for easy oil adsorption. J Colloid Interface Sci 560
19. Cheng H, Gu B, Pennefather MP, Nguyen TX, Phan-Thien N, Duong HM (2017) Cotton aerogels and cotton-cellulose aerogels from environmental waste for oil spillage cleanup. Mater Des 130
20. Liao Q, Su X, Zhu W, Hua W, Qian Z, Liu L et al (2016) Flexible and durable cellulose aerogels for highly effective oil/water separation. RSC Adv 6(68)
21. Lin R, Li A, Zheng T, Lu L, Cao Y (2015) Hydrophobic and flexible cellulose aerogel as an efficient, green and reusable oil sorbent. RSC Adv 5(100)
22. Qiao K, Tian W, Bai J, Wang L, Zhao J, Du Z et al (2019) Application of magnetic adsorbents based on iron oxide nanoparticles for oil spill remediation: a review. J Taiwan Inst Chem Eng 97
23. Peng H, Wang H, Wu J, Meng G, Wang Y, Shi Y et al (2016) Preparation of superhydrophobic magnetic cellulose sponge for removing oil from water. Ind Eng Chem Res 55(3)
24. Nechita P (2020) Use of recycled cellulose fibers to obtain sustainable products for bioeconomy applications. In: Generation, development and modifications of natural fibers
25. Nguyen ST, Feng J, Le NT, Le ATT, Hoang N, Tan VBC et al (2013) Cellulose aerogel from paper waste for crude oil spill cleaning. Ind Eng Chem Res 52(51)

26. Prathap A, Sureshan KM (2017) Organogelator–cellulose composite for practical and eco-friendly marine oil-spill recovery. Angew Chemie Int Ed 56(32)
27. Chinga-Carrasco G (2011) Cellulose fibres, nanofibrils and microfibrils: the morphological sequence of MFC components from a plant physiology and fibre technology point of view. Nanoscale Res Lett 6
28. Fürtauer S, Hassan M, Elsherbiny A, Gabal SA, Mehanny S, Abushammala H (2021) Current status of cellulosic and nanocellulosic materials for oil spill cleanup. Polymers 13
29. Abdul Khalil HPS, Davoudpour Y, Islam MN, Mustapha A, Sudesh K, Dungani R et al (2014) Production and modification of nanofibrillated cellulose using various mechanical processes: a review. Carbohydr Polym 99
30. Korhonen JT, Kettunen M, Ras RHA, Ikkala O (2011) Hydrophobic nanocellulose aerogels as floating, sustainable, reusable, and recyclable oil absorbents. ACS Appl Mater Interfaces 3(6)
31. Isogai A (2013) Wood nanocelluloses: fundamentals and applications as new bio-based nanomaterials. J Wood Sci 59:449–459
32. Harini K, Chandra Mohan C (2020) Isolation and characterization of micro and nanocrystalline cellulose fibers from the walnut shell, corncob and sugarcane bagasse. Int J Biol Macromol 163
33. Huang J, Wang S, Lyu S, Fu F (2018) Preparation of a robust cellulose nanocrystal superhydrophobic coating for self-cleaning and oil-water separation only by spraying. Ind Crops Prod 122
34. Akhlamadi G, Goharshadi EK (2021) Sustainable and superhydrophobic cellulose nanocrystal-based aerogel derived from waste tissue paper as a sorbent for efficient oil/water separation. Process Saf Environ Prot 154
35. Betlej I, Krajewski KJ, Boruszewski P, Zakaria S (2021) Bacterial cellulose-properties and its potential application. Sains Malaysiana 50(2)
36. Sai H, Fu R, Xing L, Xiang J, Li Z, Li F et al (2015) Surface modification of bacterial cellulose aerogels' web-like skeleton for oil/water separation. ACS Appl Mater Interfaces 7(13):7373–7381
37. Stanisławska A (2016) Bacterial nanocellulose as a microbiological derived nanomaterial. Adv Mater Sci 16(4)
38. Ieamviteevanich P, Palaporn D, Chanlek N, Poo-Arporn Y, Mongkolthanaruk W, Eichhorn SJ et al (2020) Carbon nanofiber aerogel/magnetic core-shell nanoparticle composites as recyclable oil sorbents. ACS Appl Nano Mater 3(4)
39. Wang F, Zhao X, Wahid F, Zhao X, Qin X, Bai H et al (2021) Sustainable, superhydrophobic membranes based on bacterial cellulose for gravity-driven oil/water separation. Carbohydr Polym 253
40. Jamalludin MR, Hubadillah SK, Harun Z, Othman MHD, Yunos MZ (2019) Novel superhydrophobic and superoleophilic sugarcane green ceramic hollow fibre membrane as hybrid oil sorbent-separator of real oil and water mixture. Mater Lett 240
41. Alaa El-Din G, Amer AA, Malsh G, Hussein M (2018) Study on the use of banana peels for oil spill removal. Alexandria Eng J 57
42. Husseien M, Amer AA, El-Maghraby A, Hamedallah N (2015) Oil spill removal from water by using corn stalk: factors affecting sorption process. Int J Environ Waste Manag 16(4)
43. Thilagavathi G, Praba Karan C, Thenmozhi R (2020) Development and investigations of kapok fiber based needle punched nonwoven as eco-friendly oil sorbent. J Nat Fibers 17(1)
44. Husseien M, Amer AA, El-Maghraby A, Taha NA (2009) Availability of barley straw application on oil spill clean up. Int J Environ Sci Technol 6(1)
45. Abdelwahab O, Nasr SM, Thabet WM (2017) Palm fibers and modified palm fibers adsorbents for different oils. Alexandria Eng J 56(4)
46. Chai W, Liu X, Zou J, Zhang X, Li B, Yin T (2015) Pomelo peel modified with acetic anhydride and styrene as new sorbents for removal of oil pollution. Carbohydr Polym 132
47. Kenes K, Yerdos O, Zulkhair M, Yerlan D (2012) Study on the effectiveness of thermally treated rice husks for petroleum adsorption. J Non Cryst Solids 358(22)

48. Zang D, Liu F, Zhang M, Gao Z, Wang C (2015) Novel superhydrophobic and superoleophilic sawdust as a selective oil sorbent for oil spill cleanup. Chem Eng Res Des 102
49. Lv E, Xia W, Tang M, Pu Y (2016) Preparation of an efficient oil-spill adsorbent based on wheat straw. BioResources 12(1)
50. Ali N, El-Harbawi M, Jabal AA, Yin CY (2012) Characteristics and oil sorption effectiveness of kapok fibre, sugarcane bagasse and rice husks: oil removal suitability matrix. Environ Technol 33(4)
51. Sivanesan I, Gopal J, Muthu M, Shin J, Oh JW (2021) Reviewing chitin/chitosan nanofibers and associated nanocomposites and their attained medical milestones. Polymers 13
52. Iber BT, Kasan NA, Torsabo D, Omuwa JW (2022) A review of various sources of chitin and chitosan in nature. J Renew Mater
53. Elieh-Ali-Komi D, Hamblin MR (2016) Chitin and chitosan: production and application of versatile biomedical nanomaterials. Int J Adv Res 4(3)
54. Rinaudo M (2006) Chitin and chitosan: properties and applications. Prog Polym Sci (Oxford) 31
55. Santos VP, Marques NSS, Maia PCSV, de Lima MAB, Franco L de O, de Campos-Takaki GM (2020) Seafood waste as attractive source of chitin and chitosan production and their applications. Int J Mol Sci 21
56. Ibrahim HM, El-Zairy EMR (2015) Chitosan as a biomaterial—structure, properties, and electrospun nanofibers. Concepts, Compd Altern Antibacterials
57. Setiati R, Siregar S, Wahyuningrum D, Rinanti A (2021) Synthesis method of chitin become chitosan polymer from shrimp shells for enhanced oil recovery. In: IOP conference series: earth and environmental science
58. Escudero-Oñate C, Martínez-Francés E (2018) A review of chitosan-based materials for the removal of organic pollution from water and bioaugmentation. In: Chitin-chitosan—myriad functionalities in science and technology
59. Kandra P, Challa MM, Kalangi Padma Jyothi H (2012) Efficient use of shrimp waste: present and future trends. Appl Microbiol Biotechnol 93
60. Zhang K, Zhang B, Song X, Liu B, Jing L, Chen B (2018) Generation of shrimp waste-based dispersant for oil spill response. Environ Sci Pollut Res 25(10)
61. Piotrowska-Kirschling A, Szelągowska-Rudzka K, Karczewski J, Brzeska J (2021) Application of shrimp waste for the synthesis of polyurethane-chitosan materials with potential use in sorption of oil micro-spills in water treatment. Sustainability 13(9)
62. Hosny R, Fathy M, Ramzi M, Abdel Moghny T, Desouky SEM, Shama SA (2016) Treatment of the oily produced water (OPW) using coagulant mixtures. Egypt J Pet 25(3)
63. Gadge KK, Bahekar A (2016) Investigation of mechanical properties of crab shell: a review. Int J Latest Trends Eng Technol 8(1)
64. Roy S, Goh KL, Verma C, Dasgupta Ghosh B, Sharma K, Maji PK (2022) A facile method for processing durable and sustainable superhydrophobic chitosan-based coatings derived from waste crab shell. ACS Sustain Chem Eng 10(14)
65. Doshi B, Repo E, Heiskanen JP, Sirviö JA, Sillanpää M (2017) Effectiveness of N,O-carboxymethyl chitosan on destabilization of Marine Diesel, Diesel and Marine-2T oil for oil spill treatment. Carbohydr Polym 167
66. Tan C, Han F, Zhang S, Li P, Shang N (2021) Novel bio-based materials and applications in antimicrobial food packaging: recent advances and future trends. Int J Mol Sci 22
67. Sajilata MG, Singhal RS, Kulkarni PR (2006) Resistant starch—a review. Compr Rev Food Sci Food Saf 5
68. Wang F, Ma R, Tian Y (2022) Superhydrophobic starch-based nanocomposite cryogel for oil removal underwater and magnetically guided oil slick cleanup. Carbohydr Polym 287
69. Wang F, Ma R, Tian Y (2022) Superhydrophobic starch-based adsorbent with honeycomb coral-like surface fabricated via facile immersion process for removing oil from water. Int J Biol Macromol 207
70. Wang F, Chang R, Ma R, Tian Y (2021) Eco-friendly and superhydrophobic nano-starch based coatings for self-cleaning application and oil-water separation. Carbohydr Polym 271

71. Pawar SN, Edgar KJ (2012) Alginate derivatization: a review of chemistry, properties and applications. Biomaterials 33
72. Yang JS, Xie YJ, He W (2011) Research progress on chemical modification of alginate: a review. Carbohydr Polym 84
73. Paques JP, Van Der Linden E, Van Rijn CJM, Sagis LMC (2014) Preparation methods of alginate nanoparticles. Adv Colloid Interface Sci 209
74. Mokhena TC, Mochane MJ, Mtibe A, John MJ, Sadiku ER, Sefadi JS (2020) Electrospun alginate nanofibers toward various applications: a review. Materials (Basel) 13(4)
75. Ayarza J, Coello Y, Nakamatsu J (2017) SEM–EDS study of ionically cross-linked alginate and alginic acid bead formation. Int J Polym Anal Char 22(1)
76. Eskhan A, Banat F (2018) Removal of oil from water by calcium alginate hydrogel modified with maleic anhydride. J Polym Environ 26(7)
77. Zhuang J, Dai J, Ghaffar SH, Yu Y, Tian Q, Fan M (2020) Development of highly efficient, renewable and durable alginate composite aerogels for oil/water separation. Surf Coatings Technol 388
78. Madhubashani AMP, Giannakoudakis DA, Amarasinghe BMWPK, Rajapaksha AU, Pradeep Kumara PBT, Triantafyllidis KS et al (2021) Propensity and appraisal of biochar performance in removal of oil spills: a comprehensive review. Environ Pollut 288
79. Gao Z, Zhang Y, Song N, Li X (2017) Biomass-derived renewable carbon materials for electrochemical energy storage. Mater Res Lett 5
80. Wang J, Wang S (2019) Preparation, modification and environmental application of biochar: a review. J Cleaner Prod 227
81. Cha JS, Park SH, Jung SC, Ryu C, Jeon JK, Shin MC et al (2016) Production and utilization of biochar: a review. J Indus Eng Chem 40
82. Song B, Cao X, Gao W, Aziz S, Gao S, Lam CH et al (2022) Preparation of nano-biochar from conventional biorefineries for high-value applications. Renew Sustain Energy Rev 157
83. Sharifi S, Behzadi S, Laurent S, Forrest ML, Stroeve P, Mahmoudi M (2012) Toxicity of nanomaterials. Chem Soc Rev 41(6)

# Biodegradable Waste Sorbents for Oil Spill Treatment

**Imtiyaz Akbar Najar, Showkat Ahmad Bhawani, Aamir Hussain Bhat, Raudhah Ahmadi, Shanshan Tang, Yunika Kirana Abdul Khalik, and Nadeem Akbar Najar**

**Abstract** The environmental and economic consequences of oil spills make them a worldwide concern. To address this issue, several commercial systems have been developed to mitigate and control oil spills. Among these systems, the utilization of biodegradable sorbents has emerged as an effective approach. These sorbents have the advantage of being capable of breaking down naturally over time, reducing the long-term impact of oil spills on the environment. Biodegradable sorbents are emerging as a sustainable and environmentally friendly solution for oil spill treatment. Derived from natural and renewable sources, these sorbents possess the ability to adsorb or absorb oil, facilitating containment and recovery efforts. This chapter provides an overview of the characteristics, applications, and benefits of biodegradable waste sorbents in oil spill treatment. This chapter highlights various types of biodegradable sorbents, including natural fiber sorbents, cellulose-based materials, agricultural by-products, mushroom spent substrate, and alginate-based materials. These sorbents exhibit high oil sorption capacities and can be deployed in oil spill response, shoreline clean-up, marsh and wetland restoration, and controlled treatment facilities. Biodegradable waste sorbents offer advantages such as reduced dependence

I. A. Najar (✉) · R. Ahmadi · Y. K. A. Khalik
Faculty of Engineering, Universiti Malaysia Sarawak, Kota Samarahan, Sarawak, Malaysia
e-mail: imtiyaznajar999@gmail.com

S. A. Bhawani
Faculty of Resource Science and Technology, Universiti Malaysia Sarawak, Kota Samarahan, Sarawak, Malaysia

A. H. Bhat
College of Applied Science and Pharmacy, University of Technology and Applied Sciences, Muscat, Oman

S. Tang
Faculty of Engineering, Anhui Sanlian University, Hefei, China

N. A. Najar
Centre for Policy Studies, Indian Institute of Technology Bombay, Mumbai, India

© The Author(s), under exclusive license to Springer Nature Singapore Pte Ltd. 2025

S. A. Bhawani et al. (eds.), *Biobased Materials and Their Composites for Oil Spill Treatment*, Composites Science and Technology,
https://doi.org/10.1007/978-981-96-6049-0_5

on non-renewable resources, sustainable waste management practices, and compatibility with environmental protection goals. However, their effectiveness and deployment strategies depend on factors such as spill characteristics, location, and regulatory guidelines. In summary, biodegradable waste sorbents provide a promising and environmentally conscious approach to oil spill treatment. Their utilization promotes sustainability, protects ecosystems, and contributes to a more resilient and greener future in oil spill response and environmental management.

**Keywords** Biodegradable · Environment · Oil spill treatment · Sorbent · Sustainable

## 1 Introduction

An oil spill refers to the accidental or deliberate release of liquid petroleum hydrocarbons, commonly known as oil, into the environment, particularly into bodies of water such as oceans, rivers, or lakes [1, 2]. Oil spills can occur during various activities related to the extraction, transportation, storage, and usage of oil, including offshore drilling, tanker accidents, pipeline leaks, and refinery or storage facility mishaps. When an oil spill happens, the oil can spread rapidly and contaminate the surrounding environment, posing significant ecological and economic risks. The severity of the impact depends on factors such as the volume and type of oil spilled, the location and sensitivity of the affected ecosystem, and the effectiveness of response measures [1].

To fully comprehend the extent to which oil spills affect marine ecosystems and their capacity to recover, it is crucial to consider an additional layer of impacts shown in Fig. 1. Oil spills have several adverse effects on the environment. The oil forms a floating layer on the water's surface, which can suffocate marine organisms by blocking sunlight and reducing oxygen levels [3, 4]. It can also coat the feathers or fur of birds and mammals, impairing their ability to fly, swim, or insulate themselves, leading to hypothermia or drowning [5, 6]. Additionally, the toxic components of oil can harm fish, shellfish, and other aquatic organisms, disrupting the food chain and causing long-term ecological damage [7]. Moreover, oil spills present a complex challenge for responders and environmental agencies. Containing and mitigating the spread of the spilled oil is a race against time, as the oil can rapidly disperse and spread, driven by wind, currents, and tides. Responders face the daunting task of employing various methods and technologies to minimize the impact on sensitive habitats, wildlife, and human health. Decisions must be made regarding the most effective strategies for containment, clean-up, and restoration, considering the unique characteristics of each spill, including the type of oil, location, weather conditions, and available resources [8].

Cleaning up an oil spill is a complex and challenging task. Response efforts typically involve containment to prevent the spread of the oil, mechanical methods such as booms and skimmers to remove the oil from the water's surface, and the use

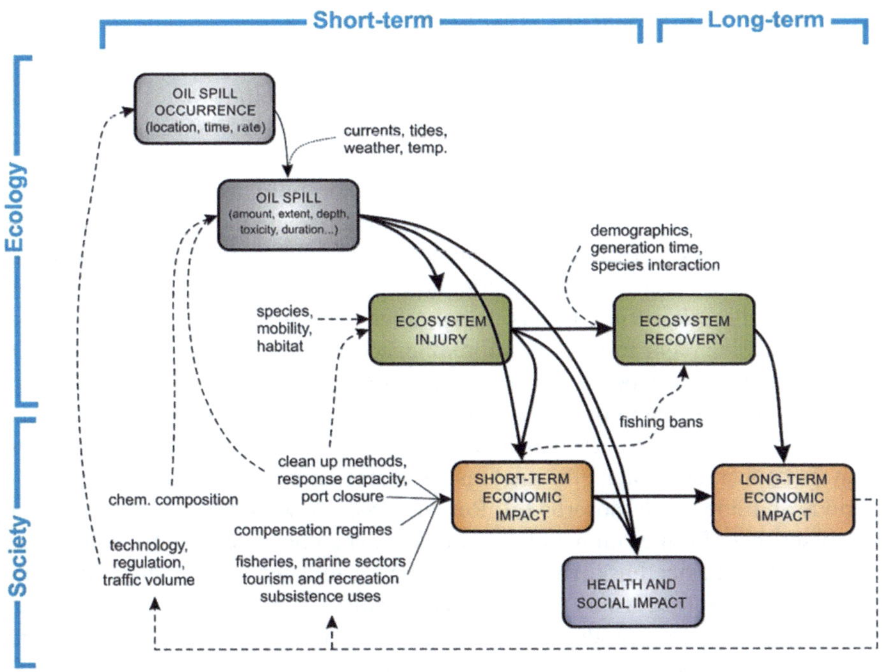


**Fig. 1** Effects of the oil spill on society and the environment [9]

of dispersants or biological agents to break down the oil. However, complete removal of spilled oil is often difficult, and the environmental impacts can persist for years or even decades [4, 10]. To prevent oil spills, stringent regulations, safety measures, and technological advancements are in place across various industries involved in oil production, transportation, and storage [11]. Additionally, emergency response plans, environmental impact assessments, and increased awareness of the risks associated with oil spills contribute to minimizing their occurrence and mitigating their effects [12, 13].

Over the years, extensive research and technological advancements have led to the development of improved oil spill response techniques. Mechanical containment methods, chemical dispersants, bioremediation approaches, and the use of innovative sorbents have all contributed to enhancing the efficiency and effectiveness of clean-up operations. However, despite these efforts, the scale and complexity of oil spills demand ongoing innovation and preparedness to minimize their devastating effects. In this context, understanding the causes, consequences, and potential solutions to oil spills becomes crucial. This knowledge empowers policymakers, environmental organizations, and industries to prioritize prevention, preparedness, and response measures. By raising awareness and fostering collaboration, it is possible to reduce the frequency and magnitude of oil spills, protect fragile ecosystems, and safeguard the livelihoods of communities worldwide.

## 2 Synthesis of Existing Research

Sorbents are highly effective in attracting oil while repelling water. They remove oil through two main processes: adsorption and absorption. In adsorption, oil sticks to the surface of the sorbent, with the thickness of the layer depending on the oil's viscosity, as thicker oils form denser layers. Absorption, however, involves capillary attraction, where oil fills the pores of the material and moves upward. Sorbents are generally classified into inorganic minerals, synthetic organics, and natural agricultural products. Sorbents from agriculture, like cotton, straw, kenaf, corn cobs, coconut shells, silkworm cocoons, kapok fibers, rice husk, hay, sawdust, bagasse, gorse, and dried palm fronds, are affordable, eco-friendly, and easy to use [14–16]. Their effectiveness depends on factors like sorption capacity, density, wettability, retention, and recyclability [17, 18]. These things are oleophilic due to their waxy surface and, when dried, are buoyant, aiding in their floating ability. Among these, straw is known for its high efficiency, as it remains on the water surface for extended periods and effectively absorbs oil. Research has shown straw's sorption capacity surpasses that of some commercially available synthetic materials [19]. Other materials like kapok, rice husk, banana trunk fiber, acetylated cotton, and cotton grass fiber have also shown strong oil sorption capabilities [20–25].

Oil and petroleum products have a significant impact on the global economy, driving various industries worldwide. However, the exploration, production, and transportation of oil pose an inevitable risk of oil spill pollution. Oil spills can cause extensive environmental and ecological damage, transcending boundaries and borders. Throughout history, oil spill accidents have occurred due to different processes involved in production and transportation. While oil spills are stochastic in nature, they typically result from accidents or intentional discharge by humans [26]. Oil slicks and the intrusion of oil into marine environments have detrimental effects on dissolved oxygen levels, negatively impacting aquatic ecosystems' well-being [4]. Birds' feathers and mammals' fur that rely on marine species for food can become coated with oil, leading to sickness or death [3, 27]. Recovering from an oil spill is a challenging and complex task. Inappropriate techniques and measures can cause further environmental damage. Effective management of resources, both in terms of materials and human resources, is crucial for developing proactive strategies to contain oil spills. The availability of financial resources to cover logistics costs, as well as trained personnel, remuneration, and support staff, is essential for mitigating the devastating effects of oil spills. The interplay between manpower and materials is evident when dealing with oil spills [28]. Any material considered for containing oil spills should possess broad applicability, be cost-effective, and possess the necessary properties to restore the natural habitat to its original or near-original condition in a timely and efficient manner.

Various methods are utilized for oil spill recovery and remediation, falling into four main categories: sorption processes, bioremediation, chemical treatment, and physical (including mechanical methods and in situ burning) [29, 30]. Physical approaches use booms, skimmers, and sorbents to contain and recover oil. These materials act as barriers, preventing the spread of oil slicks. While effective, these approaches often require substantial financial investment and logistical support. Booms, as categorized by Ventikos et al. [31], are mechanical barriers used for oil containment, with types such as fire-resistant booms, curtain booms, and fence booms effectively restricting oil spread. Skimmers, designed to lift oil from the water surface, maintain the oil's composition for potential reuse, though factors like low visibility and rough seas can limit their effectiveness. In situ burning is a cost-efficient technique but is restricted due to environmental and safety risks.

Chemical treatment involves the application of dispersants or solidifiers to remove spilled oil. This procedure is fairly straightforward and has the capacity to rapidly address extensive areas affected by pollution. Nevertheless, chemical treatment encounters certain drawbacks. Dispersants, for instance, can pose a threat to the aquatic ecosystem and often demonstrate inefficiency when dealing with highly viscous and emulsified oil. Additionally, the application of dispersants in remote areas is constrained by limited dispersion windows. The use of dispersants and solidifiers is also costly, with high initial and ongoing expenses [30]. Solidifiers, typically made from synthetic polymers, are expensive and primarily suitable for removing small, thin oil slicks [32]. Bioremediation is another important method for oil spill clean-up. It is cost-effective and generally achieves high removal rates. However, it is less effective for spills with large coherent masses and often requires a longer time for proper removal [30].

Sorbents are materials utilized for the removal and recovery of pollutants through absorption, adsorption, or a combination of both processes [33]. Adsorbents function by externally coating the surface of the sorbent material with oil, while absorption involves the incorporation of oil into the absorbent pores shown in Fig. 2. The effectiveness of sorbents is closely related to their porosity. Various organic and inorganic materials have been used as sorbents for oil spill cleanup, making them one of the most promising solutions due to their simplicity and high efficiency [34, 35]. In order for a material to be considered appropriate for use as an oil spill sorbent, it must possess certain physical properties. These properties include a natural affinity for oil (oleophilicity), resistance to water (hydrophobicity), the ability to absorb and retain large quantities of oil, the capability to biodegrade over time, and cost-effectiveness. Oil spill sorbents are typically divided into three main categories: natural organic sorbents, natural inorganic sorbents, and synthetic sorbents [36]. Natural inorganic adsorbents have limited application in spill clean-up, with adsorption capacities ranging from 4 to 20 times their weight in oil and little buoyancy [34].

Inorganic minerals such as aluminas, silicas, zeolites, clay, alumina-phosphates, and mesoporous adsorbents are among the natural inorganic sorbents used in oil spill clean-up [37]. Synthetic adsorbents, which are widely employed commercially, include polystyrene, polypropylene, and polyester foams. These synthetic materials exhibit high hydrophobic and oleophilic properties and can sorb up to 70 times their weight in oil. They can also be reused multiple times after recovery [34]. However, conventional sorbents are expensive, non-biodegradable, and not environmentally friendly. Due to their drawbacks, conventional sorbents are becoming less favoured,

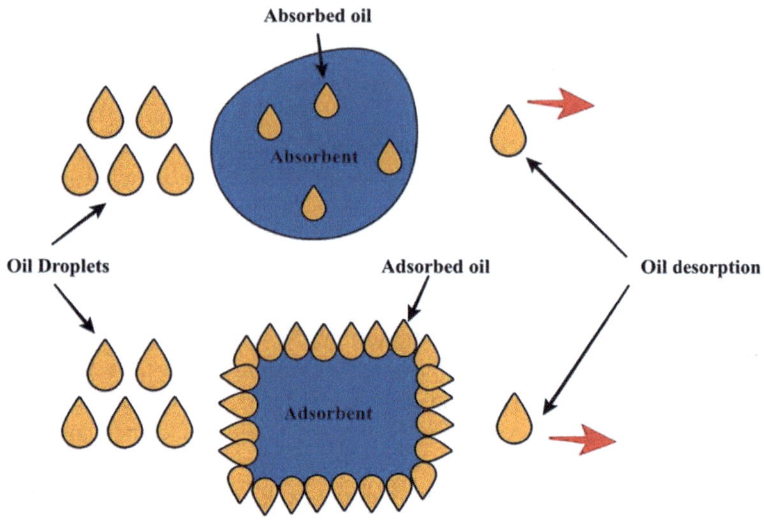


**Fig. 2** Oil absorption and adsorption processes

as they have limited oil sorption and retention capabilities, poor biodegradability, and challenging recovery processes [35].

To address these issues, researchers, scientists, and engineers have turned their attention to non-conventional materials for oil spill remediation. They have explored the use of agricultural and industrial waste biopolymers as alternative sorbents. These non-conventional sorbents are biodegradable, abundantly available in nature, and cost-effective. In particular, cellulose and keratin-based biopolymeric sorbents have gained significant attention. This comprehensive review focuses on these non-conventional sorbents in their natural and modified forms, highlighting their characteristics, mechanisms, and oil sorption capacities. These materials offer promising solutions for oil spill clean-up and oil–water separation.

## 3 Sorption Mechanisms and Factors Affecting Oil Sorption

Sorption of oil to sorbents can occur through absorption, adsorption, or a combination of both methods. Absorption involves the oil penetrating the pore spaces of the sorbents at an intermolecular level. The sorbents' ability to absorb oil is primarily influenced by their porosity and can be influenced by the oil's characteristics, such as viscosity, stickiness, and adhesion properties. The transport of oil into the absorbent material's voids is facilitated by capillary forces, which can be intensified by pressure and gravitational forces. Absorption which occurs when the sorbent material imbibes or soaks up the oil into its structure. Factors affecting absorption include: Hydrophilicity/Hydrophobicity—sorbents with hydrophobic properties tend to have a higher affinity for oil, facilitating absorption. Conversely, hydrophilic sorbents may absorb less oil due to their preference for water. Porosity—highly porous sorbent materials can absorb oil into their interconnected pore structure, effectively trapping the oil within the material. Bulk density—the density of the sorbent material affects its ability to absorb oil. Lower bulk density allows for greater oil uptake.

On the other hand, in adsorption, the oil is accumulated or retained on the sorbent's surfaces without infiltrating the material itself [33]. This process is driven by the attraction between the outer surface of the sorbents and the oil, which is controlled by various physical and chemical interactions such as Van der Waals forces, hydrogen bonding, polarity, steric interactions, and hydrophobicity [38]. Adsorption, which occurs when oil molecules adhere to the surface of the sorbent material. Factors influencing adsorption include-

- Surface area: sorbents with a higher surface area provide more sites for oil molecules to attach, leading to increased sorption capacity.
- Porosity—the presence of pores or cavities within the sorbent material enhances its adsorption capability by creating additional surface area and increasing accessibility for oil molecules.
- Surface chemistry: The surface chemistry of the sorbent material plays a crucial role in attracting and retaining oil molecules. Surface properties, such as polarity, hydrophobicity, and functional groups, can influence the sorption capacity.

The sorption process takes place in three distinct steps: Firstly, the oil diffuses to the surface of the sorbent; secondly, capillary action traps the oil within the sorbent's structure; and finally, the oil forms agglomerates within the rough and porous structure of the sorbent [39]. When biopolymeric sorbents are exposed to oil, it is believed that both absorption and adsorption mechanisms occur. The oil rapidly accumulates on the sorbent's surface through lipophilic interactions or coalescence, while absorption takes place as the oil moves into the pores of the sorbent materials through capillary action [40, 41]. Several factors influence the oil sorption capacities of biopolymers, including the packing and bulk density of the materials, void volume (porosity), surface morphology, selectivity, and oil retention properties of the biopolymers. Studies have demonstrated that the presence of functional groups such as hydroxyl, carboxylic, and amino acid groups can impact the sorption properties of biopolymers [4, 8, 34]. Factors affecting oil sorption include:

I. **Hydrodynamic conditions**—the movement and turbulence of water can affect the sorption process. Higher flow rates or increased wave action can hinder sorption by reducing the contact time between the sorbent and the oil. Conversely, calmer conditions facilitate better sorption efficiency.
II. **Oil properties**—the characteristics of the spilled oil itself influence sorption. Factors include oil viscosity, density, volatility, and composition. Oils with higher viscosity or lower volatility tend to be more readily sorbed by waste sorbents.
III. **Sorbent properties**—the specific properties of the biodegradable sorbent material impact its oil sorption effectiveness. These include: Material structure—the

physical structure, such as fiber length, size, and arrangement, affects sorption efficiency. Materials with a higher aspect ratio, such as long fibers, often exhibit enhanced sorption capacity.

IV. **Hydrophobicity/Hydrophilicity**—sorbents with hydrophobic properties have a greater affinity for oil and tend to exhibit improved sorption performance. Sorbent stability—the stability and structural integrity of the sorbent material under the given environmental conditions influence its ability to effectively sorb and retain oil.

V. **Environmental Factors**—environmental conditions, such as temperature, pH, salinity, and presence of other contaminants, can affect sorption efficiency. Extreme temperatures, high salinity levels, or the presence of certain chemicals may hinder sorption. It is important to consider these factors when selecting biodegradable sorbents for oil spill treatment to ensure optimal sorption performance and maximize the effectiveness of the clean-up operation.

## 4 Biodegradable Waste Sorbents as a Treatment Technique

Biodegradable waste sorbents offer a sustainable and environmentally friendly approach to oil spill treatment. These materials, derived from natural and renewable sources, have the ability to adsorb or absorb oil, aiding in containment and removal efforts. Let us explore in detail the characteristics and benefits of biodegradable waste sorbents for oil spill treatment.

I. **Natural fiber Sorbents**: are derived from organic materials such as straw, hay, coconut coir, peat moss, and sawdust. These materials possess a high affinity for oil, allowing them to effectively adsorb and retain it. They can be readily deployed onto the water's surface to soak up floating oil or placed on shorelines to capture oil that has reached the land. Natural fiber sorbents are biodegradable, meaning they can naturally break down over time without leaving behind harmful residues. This makes them a desirable option for oil spill treatment, as they minimize the long-term impact on the environment [19–21].

II. **Cellulose-based Sorbents**: are derived from plant sources such as cotton, paper, or cellulose pellets. These materials are effective in oil sorption due to their porous structure and high surface area. They can rapidly absorb oil, helping to contain and mitigate the spread of the spill. Cellulose-based sorbents are biodegradable and can be easily disposed of or processed without causing harm to the environment. They are often used in various forms, such as pads, booms, or mats, for efficient oil sorption [24, 30, 35].

III. **Agricultural by-products**: agricultural waste materials can be repurposed as oil sorbents, providing an environmentally friendly solution. Waste materials like rice husks, corn cobs, and wheat straw, which are abundant by-products of agricultural processes, have been found to be effective sorbents for oil spills. These materials have a natural affinity for oil and can be used to absorb and remove it from the water's surface. Once saturated, the oil-laden sorbents can be composted or recycled as soil amendments, completing the cycle of sustainability [15, 20, 22, 42].

IV. **Mushroom spent substrate**: mushroom farming generates a significant amount of waste known as mushroom spent substrate (MSS). This waste material, which consists of lignocellulosic materials, has been found to have excellent oil sorption capabilities. MSS can adsorb and retain oil effectively, making it a potential biodegradable sorbent for oil spill treatment. A notable advantage of using MSS is the potential for reusability. After oil sorption, the spent substrate can be processed and used as a substrate for mushroom cultivation, offering an innovative and sustainable approach to waste management [21, 43].

V. **Alginate-based sorbents**: Alginate, a natural biopolymer derived from seaweed or algae, can be processed into various forms, such as beads or mats, for use as oil sorbents. Alginate-based sorbents have high oil absorption capacity and can be deployed to capture and remove oil from the water's surface. They are biodegradable and can be disposed of without causing harm to the environment. The utilization of alginate-based sorbents contributes to the sustainable management of oil spills and promotes the use of renewable resources [44].

Biodegradable waste sorbents offer several advantages over traditional sorbents made from synthetic materials. They are derived from renewable sources, reducing the dependence on non-renewable resources. Their biodegradable nature ensures minimal long-term environmental impact, as they break down naturally over time. Additionally, these sorbents can often be processed or reused in an environmentally friendly manner, completing the cycle of sustainability. However, it is important to note that the effectiveness of biodegradable waste sorbents may vary depending on factors such as sorbent composition, oil type, environmental conditions, and the specific application method. Proper evaluation and selection of the appropriate sorbent based on the spill scenario and site conditions are crucial to ensure optimal oil sorption performance. Overall, biodegradable waste sorbents offer a promising and environmentally conscious solution for oil spill clean-up and pollution control, aligning with the principles of sustainability and resource conservation.

## 5 Applications

Biodegradable waste sorbents have found numerous applications in oil spill treatment due to their effectiveness, environmental compatibility, and sustainable nature. Here are some key applications of biodegradable waste sorbents:

I. **Surface Oil Recovery**: Biodegradable waste sorbents, such as natural fiber mats, pads, or booms, can be deployed on the water's surface to absorb or adsorb the spilled oil. These sorbents have a high oil absorption capacity and can efficiently capture and retain the oil, preventing its further spread. They are particularly useful in calm waters or during early stages of oil spill response when the oil is still primarily floating on the surface [43, 45].

II. **Shoreline Protection**: Biodegradable sorbents can be utilized to protect shorelines and sensitive habitats from oil contamination. When oil reaches the coastline, biodegradable sorbents, such as natural fiber rolls or straw bales, can be strategically placed to create barriers and absorb or adsorb the oil before it infiltrates the shoreline or reaches critical ecosystems. This helps minimize the ecological impact and simplifies subsequent clean-up efforts [46].

III. **Marsh and Wetland Restoration**: Marshes and wetlands are delicate ecosystems that are highly susceptible to the adverse effects of oil spills. Biodegradable waste sorbents, such as agricultural by-products or natural fiber mats, can be employed to absorb or adsorb oil in these sensitive environments. By reducing the oil contamination, biodegradable sorbents aid in the restoration of marshes and wetlands, allowing for the recovery of their biodiversity and ecological functions [3].

IV. **Inland Water Treatment**: Biodegradable sorbents are effective in treating oil spills in inland water bodies, such as lakes, rivers, and ponds. Natural fiber sorbents or cellulose-based materials can be used to absorb or adsorb the oil, allowing for its removal from the water. These sorbents can be strategically placed in areas where the oil is concentrated or carried by currents, facilitating its containment and recovery.

V. **Clean-up of Sensitive Habitats**: Biodegradable waste sorbents offer a gentle and non-intrusive approach to cleaning up oil spills in environmentally sensitive areas. When oil affects delicate habitats, such as coral reefs, seagrass beds, or mangrove forests, biodegradable sorbents can be used to selectively remove the oil without causing further damage. These sorbents can be applied manually, minimizing disturbance to the fragile ecosystems and promoting their recovery.

VI. **Post-spill Remediation**: After the initial containment and recovery efforts, biodegradable waste sorbents can aid in the post-spill remediation process. They can be used to collect residual oil or oil that has undergone weathering, allowing for a more thorough clean-up. These sorbents help in removing lingering oil from the environment, promoting the restoration of affected areas.

It is important to consider the specific characteristics of the spilled oil, environmental conditions, and regulatory guidelines when selecting and implementing biodegradable waste sorbents for oil spill treatment. The proper disposal or treatment of the oil-laden sorbents after use is also essential to ensure minimal environmental impact.

## 6 Advantages

Biodegradable waste sorbents offer several advantages for oil spill treatment compared to traditional sorbents made from synthetic materials. Here are some key advantages:

I. **Sustainability**: Biodegradable waste sorbents are derived from natural and renewable sources, such as plant fibers, agricultural by-products, or mushroom spent substrate. Using these materials reduces reliance on non-renewable resources and promotes sustainable practices in oil spill response and clean-up.

II. **Environmental Friendliness**: Biodegradable sorbents break down naturally over time, minimizing their long-term impact on the environment. Unlike synthetic sorbents, which can persist in the environment for extended periods, biodegradable sorbents degrade into harmless by-products, reducing waste accumulation.

III. **Safe for Wildlife**: Biodegradable sorbents pose a lower risk to wildlife. Synthetic sorbents can be hazardous if ingested by animals, leading to internal blockages or other health issues. Biodegradable sorbents, on the other hand, are often made from natural fibers that are less likely to cause harm if accidentally consumed by wildlife.

IV. **Ease of Disposal**: After oil sorption, biodegradable sorbents can be disposed of or processed without causing harm to the environment. Depending on the specific sorbent material, disposal options may include composting, controlled incineration, or recycling as soil amendments. This simplifies the clean-up process and minimizes the need for long-term storage or specialized waste management.

V. **Reusability**: Some biodegradable sorbents, such as mushroom spent substrate, offer the potential for reusability. After oil sorption, the spent substrate can be processed and used as a substrate for mushroom cultivation or other applications, providing an innovative and sustainable solution to waste management.

VI. **High Sorption Capacity**: Biodegradable sorbents can have a high oil sorption capacity, effectively adsorbing or absorbing oil from the water's surface. Materials like natural fibers, cellulose-based sorbents, or alginate-based sorbents have a porous structure or high surface area, allowing them to rapidly capture and retain oil.

VII. **Versatility**: Biodegradable sorbents can be manufactured in various forms, such as pads, booms, blankets, or beads, making them adaptable to different oil spill scenarios. They can be easily deployed on water surfaces, shorelines, or other affected areas, providing flexibility in containment and clean-up operations.

VIII. **Cost-Effectiveness**: Biodegradable waste sorbents can often be produced at a lower cost compared to synthetic sorbents, as they utilize readily available waste materials or natural fibers. This cost advantage makes them an attractive option for oil spill response, especially in large-scale clean-up operations.

It is important to consider the specific characteristics and limitations of different biodegradable sorbents when selecting the appropriate material for a particular oil spill situation. Factors such as sorption capacity, buoyancy, durability, and compatibility with other clean-up methods should be evaluated to ensure effective and efficient oil spill treatment.

## 7 Conclusion

In conclusion, biodegradable waste sorbents offer a encouraging and environmentally friendly tactic to oil spill treatment. Resulting from natural and renewable sources, these sorbents determine effective oil adsorption and absorption capabilities while diminishing the long-term effect on the environment. Their applications range from oil spill response and shoreline clean-up to marsh and wetland restoration, contributing to the protection of delicate ecosystems. Biodegradable waste sorbents, such as natural fibers, cellulose-based materials, agricultural by-products, mushroom spent substrate, and alginate-based materials, have shown their potential in containing, recovering, and mitigating oil spills. These sorbents not only aid in the clean-up and recovery efforts but also align with sustainable waste management practices. After use, they can be composted, recycled, or processed without leaving harmful residues, completing the cycle of sustainability.

The use of biodegradable waste sorbents in oil spill remediation is indicative of a rising recognition of the need for ecologically friendly solutions. Oil spill management becomes more robust and sustainable when sorbents prioritise renewable resources and reduce long-term effects on communities and ecosystems. Nonetheless, to improve the effectiveness and performance of biodegradable sorbents even further, continuous research, development, and innovation are necessary. Sustained endeavours to enhance sorption capability, durability, and deployment tactics will augment their broader acceptance and efficacy in forthcoming oil spill remediation situations. Biodegradable waste sorbents present a viable option for sustainable oil spill treatment, safeguarding delicate ecosystems, wildlife, and human communities while promoting a greener and more resilient future, as the world works to prevent and lessen the environmental impact of oil spills.

## References

1. Doshi B, Sillanpää M, Kalliola S (2018) A review of bio-based materials for oil spill treatment. Water Res 135:262–277
2. Idris J, Eyu GD, Mansor AM, Ahmad Z, Chukwuekezie CS (2014) A preliminary study of biodegradable waste as sorbent material for oil-spill cleanup. Sci World J 2014
3. Beyer J, Trannum HC, Bakke T, Hodson PV, Collier TK (2016) Environmental effects of the Deepwater Horizon oil spill: a review. Mar Pollut Bull 110(1):28–51. Available from: https://doi.org/10.1016/j.marpolbul.2016.06.027
4. Ifelebuegu AO, Ukpebor JE, Ahukannah AU, Nnadi EO, Theophilus SC (2017) Environmental effects of crude oil spill on the physicochemical and hydrobiological characteristics of the Nun River, Niger Delta. Environ Monit Assess 189(4)
5. Drahor MG, Göktürkler G, Berge MA, Kurtulmuş TÖ (2006) Application of electrical resistivity tomography technique for investigation of landslides: a case from Turkey. Environ Geol 50(2):147–155
6. Dezert T, Fargier Y, Palma Lopes S, Côte P (2019) Geophysical and geotechnical methods for fluvial levee investigation: a review. Eng Geol 260:105206. Available from: https://doi.org/10.1016/j.enggeo.2019.105206

7. Hoang AT, Nguyen XP, Duong XQ, Huynh TT (2021) Sorbent-based devices for the removal of spilled oil from water: a review. Environ Sci Pollut Res 28(23):28876–28910
8. Wahi R, Chuah LA, Choong TSY, Ngaini Z, Nourouzi MM (2013) Oil removal from aqueous state by natural fibrous sorbent: an overview. Sep Purif Technol 113:51–63. Available from: https://doi.org/10.1016/j.seppur.2013.04.015
9. Chang SE, Stone J, Demes K, Piscitelli M (2014) Consequences of oil spills: a review and framework for informing planning. Ecol Soc 19(2)
10. Fouad RR, Aljohani HA, Shoueir KR (2016) Biocompatible poly(vinyl alcohol) nanoparticle-based binary blends for oil spill control. Mar Pollut Bull 112(1–2):46–52. Available from: https://doi.org/10.1016/j.marpolbul.2016.08.046
11. Dave D, Ghaly AE (2011) Remediation technologies for marine oil spills: a critical review and comparative analysis. Am J Environ Sci 7(5):424–440
12. Najar IA, Ahmadi RB, Jamian MAH, Hamza HB, Ahmad A, Sin CH (2022) Site-specific ground response analysis using the geotechnical dataset in moderate seismicity region. Int J Mech 16(1):37–45
13. Ahmadi R, Najar IA, Abdullahi AF, Norazzlina M, Hamza H, Najar NA (2020) Computational investigation of soil liquefaction susceptibility based on standard penetration test value of Miri District (Sarawak, Malaysia). Int J Adv Sci Technol 29(7):2735–2748
14. Johnson RF, Manjrekar TG, Halligan JE (1973) Removal of oil from water surfaces by sorption on unstructured fibers. Environ Sci Technol 7(5):439–443
15. Tsai WT, Chang CY, Wang SY, Chang CF, Chien SF, Sun HF (2001) Utilization of agricultural waste corn cob for the preparation of carbon adsorbent. J Environ Sci Heal Part B Pestic Food Contam Agric Wastes 36(5):677–686
16. Anthony WS (1994) Absorption of oil with cotton products and kenaf. Appl Eng Agric 10(3):357–361
17. Sun XF, Sun RC, Sun JX (2003) A convenient acetylation of sugarcane bagasse using NBS as a catalyst for the preparation of oil sorption-active materials. J Mater Sci 38(19):3915–3923
18. Lehr WE (1974) Containment and recovery devices for oil spill cleanup operations. JPT, J Pet Technol 26:375–380
19. Choi HM (1996) Needlepunched cotton nonwovens and other natural fibers as oil cleanup sorbents. J Environ Sci Heal Part A Toxic/Hazard Subst Environ Eng 31(6):1441–1457
20. Lim TT, Huang X (2007) Evaluation of kapok (*Ceiba pentandra* (L.) Gaertn.) as a natural hollow hydrophobic-oleophilic fibrous sorbent for oil spill cleanup. Chemosphere 66(5):955–963
21. Abdullah MA, Rahmah AU, Man Z (2010) Physicochemical and sorption characteristics of Malaysian *Ceiba pentandra* (L.) Gaertn. as a natural oil sorbent. J Hazard Mater 177(1–3):683–691. Available from: https://doi.org/10.1016/j.jhazmat.2009.12.085
22. Kumagai S, Noguchi Y, Kurimoto Y, Takeda K (2007) Oil adsorbent produced by the carbonization of rice husks. Waste Manag 27(4):554–561
23. Sathasivam K, Mas Haris MRH (2010) Adsorption kinetics and capacity of fatty acid-modified banana trunk fibers for oil in water. Water Air Soil Pollut 213(1–4):413–423
24. Adebajo MO, Frost RL, Kloprogge JT, Kokot S (2006) Raman spectroscopic investigation of acetylation of raw cotton. Spectrochim Acta Part A Mol Biomol Spectrosc 64(2):448–453
25. Suni S, Kosunen AL, Hautala M, Pasila A, Romantschuk M (2004) Use of a by-product of peat excavation, cotton grass fibre, as a sorbent for oil-spills. Mar Pollut Bull 49(11–12):916–921
26. Etkin DS (2004) Twenty-year trend analysis of oil spills in EPA jurisdiction. In: Proceedings of the 5th biennial freshwater spills symposium [Internet], pp 1–15. Available from: http://environmental-research.com/erc_papers/ERC_paper_15.pdf
27. Paul JH, Hollander D, Coble P, Daly KL, Murasko S, English D et al (2013) Toxicity and mutagenicity of gulf of Mexico waters during and after the Deepwater Horizon oil spill. Environ Sci Technol 47(17):9651–9659
28. Zhong Z, You F (2011) Oil spill response planning with consideration of physicochemical evolution of the oil slick: a multiobjective optimization approach. Comput Chem Eng 35(8):1614–1630. Available from: https://doi.org/10.1016/j.compchemeng.2011.01.009

29. Angelova D, Uzunov I, Uzunova S, Gigova A, Minchev L (2011) Kinetics of oil and oil products adsorption by carbonized rice husks. Chem Eng J 172(1):306–311. Available from: https://doi.org/10.1016/j.cej.2011.05.114
30. Liu H, Geng B, Chen Y, Wang H (2017) Review on the aerogel-type oil sorbents derived from nanocellulose. ACS Sustain Chem Eng 5(1):49–66
31. Ventikos NP, Vergetis E, Psaraftis HN, Triantafyllou G (2004) A high-level synthesis of oil spill response equipment and countermeasures. J Hazard Mater 107(1–2):51–58
32. Fingas M (2011) Introduction to oil chemistry and properties in oil spill science and technology, pp 713–733
33. Praba Karana C, Rengasamy RS, Das D (2011) Oil spill cleanup by structured fibre assembly. Indian J Fibre Text Res 36(2):190–200
34. Abdel-Moghny T, Keshawy M (2014) An overview on the treatment of oil spill using sorbent materials: synthetic and natural oil sorbent. LAP Lambert Academic Publishing, pp 1–35
35. Ifelebuegu AO, Johnson A (2017) Nonconventional low-cost cellulose- and keratin-based biopolymeric sorbents for oil/water separation and spill cleanup: a review. Crit Rev Environ Sci Technol 47(11):964–1001
36. Zhu L, Wang Y, Wang Y, You L, Shen X, Li S (2017) An environmentally friendly carbon aerogels derived from waste pomelo peels for the removal of organic pollutants/oils. Microporous Mesoporous Mater 241:285–292. Available from: https://doi.org/10.1016/j.micromeso.2016.12.033
37. Knaebel KS (2011) Adsorbent selection
38. Sabir S (2015) Approach of cost-effective adsorbents for oil removal from oily water. Crit Rev Environ Sci Technol 45(17):1916–1945
39. Tansel B, Pascual B (2011) Removal of emulsified fuel oils from brackish and pond water by dissolved air flotation with and without polyelectrolyte use: pilot-scale investigation for estuarine and near shore applications. Chemosphere 85(7):1182–1186. Available from: https://doi.org/10.1016/j.chemosphere.2011.07.006
40. Inagaki M, Kawahara A, Nishi Y, Iwashita N (2002) Heavy oil sorption and recovery by using carbon fiber felts. Carbon N Y 40(9):1487–1492
41. Srinivasan A, Viraraghavan T (2008) Removal of oil by walnut shell media. Bioresour Technol 99(17):8217–8220
42. Azubuike CC, Chikere CB, Okpokwasili GC (2016) Bioremediation techniques–classification based on site of application: principles, advantages, limitations and prospects. World J Microbiol Biotechnol 32(11)
43. Tijani MM, Aqsha A, Mahinpey N (2016) Development of oil-spill sorbent from straw biomass waste: experiments and modeling studies. J Environ Manage 171:166–176. Available from: https://doi.org/10.1016/j.jenvman.2016.02.010
44. Das N, Chandran P (2011) Microbial degradation of petroleum hydrocarbon contaminants: an overview. Biotechnol Res Int 2011:1–13
45. Tuan Hoang A, Viet Pham V, Nam Nguyen D (2018) A report of oil spill recovery technologies. Int J Appl Eng Res 13(7):4915–4928. Available from: http://www.ripublication.com
46. Barron MG (2012) Ecological impacts of the Deepwater Horizon oil spill: implications for immunotoxicity. Toxicol Pathol 40(2):315–320

# Biomass Sorbents for Oil Spill Treatment

**Utsav Garg, Zubia Khan, and Yasser Azim**

**Abstract** The study offers a comprehensive overview of biosorption technology's application in oil spill recovery and oil–water separation, drawing attention from academia, research, and industry for its versatile potential. In this context, the focus is directed towards non-conventional, cost-effective adsorbents, primarily cellulose and keratin-based materials, which exhibit promising attributes for effective oil spill cleanup. The study arrives at several key insights: Firstly, unconventional adsorbents, owing to their natural abundance, biodegradability, and economic viability, outperform traditional counterparts in oil spill recovery efforts. Secondly, the optimal oil spill sorbent necessitates tailored qualities such as lipophilicity, hydrophobicity, biodegradability, and exceptional absorption and retention capabilities. To address the limitations of poor oleophilic and hydrophobic traits of sorbents, various mechanical, thermal, and chemical modification techniques are explored and found to be effective. The sorption capacity is revealed to be a multifaceted interplay between the sorbent's attributes, spilled oil characteristics, and the surrounding marine environment. Notably, economic analyses emerge as a critical factor in evaluating the practicality of biosorption technology for large-scale industrial applications. Looking ahead, the study underscores the importance of future research in refining chemical modifications to optimize biosorbents' oil absorption capabilities while mitigating ecological impacts. In essence, this study underscores biosorption technology's considerable promise and the persistent need for innovative research to harness its full potential in addressing real-world oil spill challenges within a framework of environmental sustainability.

U. Garg · Z. Khan · Y. Azim (✉)
Department of Applied Chemistry, Faculty of Engineering and Technology, Zakir Husain College of Engineering and Technology, Aligarh Muslim University, Aligarh, Uttar Pradesh, India
e-mail: yasser.ac@amu.ac.in; yasser.azim@gmail.com

© The Author(s), under exclusive license to Springer Nature Singapore Pte Ltd. 2025 
S. A. Bhawani et al. (eds.), *Biobased Materials and Their Composites for Oil Spill Treatment*, Composites Science and Technology,
https://doi.org/10.1007/978-981-96-6049-0_6

# 1 Introduction

In the current scenario, the clean-up of water resources is encountering a notable obstacle presented by the contamination of oil, which arises from activities such as shipping, industrial drainage, disposal, and deliberate oil releases [1]. These oil discharges encompass both accidental and deliberate emissions of petroleum hydrocarbons into the surroundings, particularly within aquatic ecosystems (Fig. 1). Within marine environments, oil spills have the potential to be carried by wind currents, initiating processes like evaporation, the creation of a thin surface layer on the water, dispersion throughout the water body, sinking, and eventual accumulation within sediments [2, 3]. Various factors like salinity, wave activity, and temperature can impact the velocity and decomposition of the spilled oil [4]. Several instances of oil spill catastrophes, notably the 1991 Kuwait oil spill (240–336 million gallons), the 2010 Deepwater Horizon spill (4.9 million barrels), and the more recent East China Sea oil spill in 2018 (138 thousand tonnes), have yielded grave repercussions. Effectively managing and recovering from such extensive oil spills necessitates substantial financial investments and a considerable amount of time [5]. Marine creatures and seabirds face heightened susceptibility to the repercussions of oil spills due to the tendency of most oils to remain afloat on the water's surface [6]. Nevertheless, should the oil manage to reach the land, it can result in disastrous outcomes for terrestrial fauna as well [7]. Beyond the immediate physical afflictions suffered by those exposed to oil spills, such as respiratory problems, skin inflammation, and ocular discomfort, there are enduring ecological ramifications. These encompass dwindling populations of aquatic life and wildlife, devastation of plant life, and pollution of water reservoirs.

The process of recuperating from an oil spill is intricate and demanding, with the potential for substantial ecological harm if incorrect techniques and protocols are employed. Creating preemptive strategies to manage oil spills entails tackling challenges related to resource management, encompassing both materials and human assets. Any substance employed to counteract an oil release must not only be economically viable but must also exhibit the essential attributes to effectively reinstate the natural environment to its original or nearly original state [8]. A variety of technologies have been harnessed to address the remediation and restoration of oil spills. The

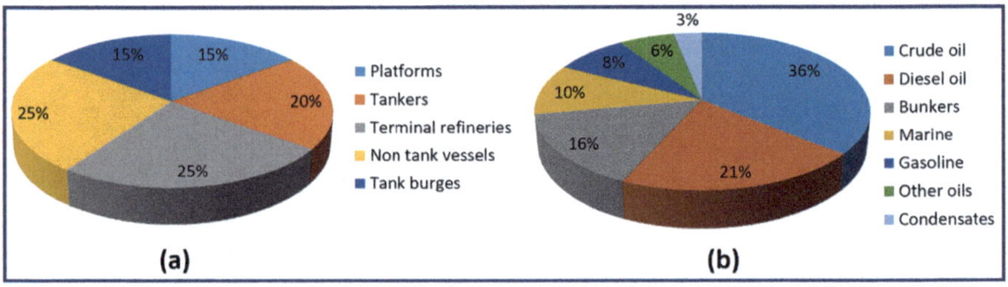


**Fig. 1** **a** Percentage (%) of origin of oil spills in the water. **b** Percentage (%) of oil types

selection of whether to employ physical, chemical, or biological approaches for the recovery and treatment of oil spills hinges primarily on two key factors: the composition of the oil and the scale of the spill itself [9] (Fig. 2). Chemical dispersants and solidifiers are frequently utilized by applying them onto the surface of an oil spill or slick. The objective is to disintegrate large oil droplets into smaller ones or solidify the spilled oil, thereby facilitating the recovery process [10, 11]. Nevertheless, the impracticality of recovering certain types of oil, especially those with high viscosity, and the potential dispersion of chemicals into marine environments are recognized as notable drawbacks of the chemical method. Moreover, the existing techniques employed for recovering oil spills frequently encounter limitations concerning their efficacy and overall effectiveness.

Moreover, the presently available dispersants and chemicals are recognized for their chemical stability, impeding biodegradation and extending their persistence in both seawater and the marine ecosystem [12]. Consequently, the demand is increasing for cost-effective, biodegradable, uncomplicated, and environmentally conscious solutions to address oil spills and chemical leaks. While biological approaches have gained attention for their ecological benefits, their reliance on microbes or bacteria for oil biodegradation often comes with a high cost and time investment [13, 14].

Amid the imperative that the employed methods be rapid, easily accessible, and user-friendly, physical pre-treatment techniques remain the most promising choice due to their remarkable recovery capabilities in comparison to other methods.

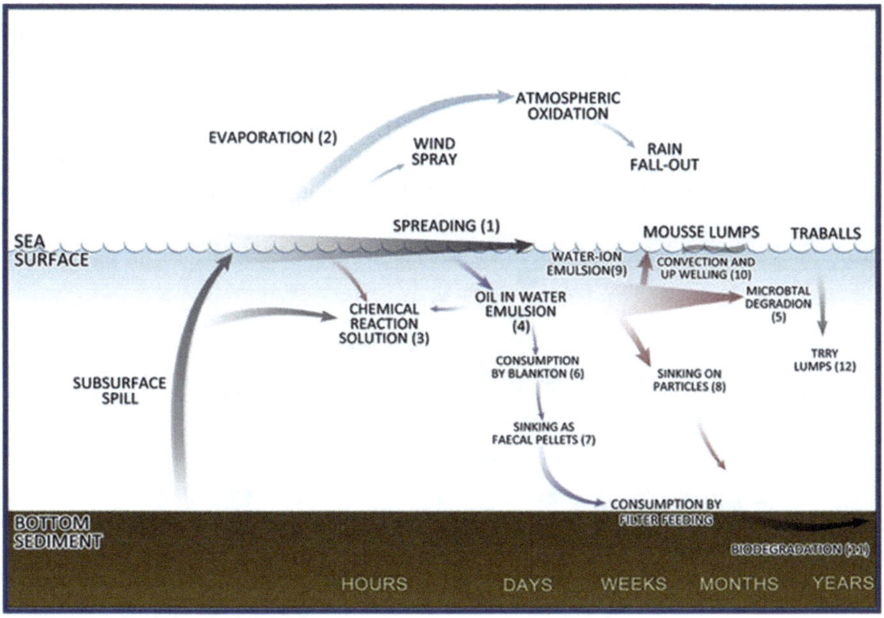


**Fig. 2** Physical, chemical and biological processes changing properties of oil slicks. Adapted with permission from Ref. [12]. Copyright 2012, Elsevier

Among the prevailing physical–mechanical methods adopted for the management and recovery of oil spills are booms, skimmers, in-situ burning, and the application of sorbent materials [15]. Nonetheless, both boom and skimming techniques tend to be ineffective and time-intensive due to the erratic nature of wind directions and wave heights.

In-situ burning approaches might be more efficient than booms or skimmers, yet they necessitate a considerable amount of time to incinerate the spilled oil and result in the release of substantial pollutants like $CO_2$, $SO_2$, PM, and VOCs [16]. Leveraging sorbents for oil spill recovery stands out as a preferred option owing to their hydrophobicity, oleophilic properties, absorption rate and capacity, oil retention, durability, buoyancy, and reusability [17, 18]. Sorbents possessing these advantageous characteristics are deemed well-suited for lifting spilled oil from the water's surface. A pivotal attribute influencing the selection of diverse sorbent types for oil recovery is their capacity to transform spilled oil from a liquid phase to a semi-solid or solid state [19].

Sorbents, substances utilized for the absorption or recovery of pollutants, function through processes of absorption, adsorption, or a combination of both [20]. Adsorbents coat the sorbent surface with oil, while absorption involves the incorporation of oil into the pores of the absorbent material (Fig. 3). Consequently, the efficacy of sorbents is contingent on their porosity. In the context of oil spill remediation, a range of organic and inorganic materials have found use as sorbents due to their user-friendliness and high efficiency, positioning them as highly promising alternatives [21, 22]. For a substance to be effective in oil spill sorption, it must possess specific physical attributes including lipophilicity, hydrophobicity, substantial absorption and retention capacity, biodegradability, and cost-effectiveness.

Oil spill sorbents are typically classified into three categories: natural organic, natural inorganic, and synthetic sorbents [23]. Natural inorganic sorbents play a limited role in spill clean-up. Although they exhibit oil-absorbing capabilities ranging from four to twenty times their weight, their buoyancy is low [22]. This group

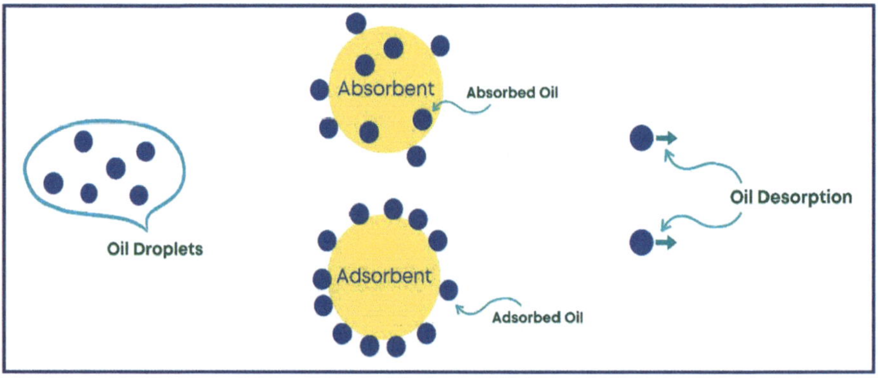


**Fig. 3** Oil absorption and adsorption processes

includes minerals like aluminas, silicas, zeolites, clays, alumina-phosphates, and mesoporous adsorbents. Conversely, synthetic sorbents are extensively employed commercially for oil spill clean-up. Common examples of synthetic adsorbents are polystyrene, polypropylene, and polyester foams. These synthetic materials possess high hydrophobic and oleophilic properties, allowing them to absorb up to seventy times their weight in oil and be reusable after recovery [22].

However, traditional sorbents used in oil spill clean-up tend to be costly, non-biodegradable, and environmentally unfriendly. Their elevated expenses, limited oil sorption and retention efficiency, and lack of biodegradability and reusability have diminished their appeal. Consequently, researchers have begun exploring the potential of biopolymers derived from agricultural and industrial waste as unconventional materials for oil spill recovery. These biopolymers are biodegradable, abundant in nature, cost-effective, and amenable to integration in innovative oil spill recovery processes. This chapter presents a comprehensive examination of low-cost cellulose and keratin-based biopolymeric sorbents, non-conventional materials employed in spill clean-up and oil–water separation in both their original and modified forms. The chapter underscores their characteristics, mechanisms, and oil sorption capacities.

## 2 Sorption Mechanisms

Sorbents exhibit pronounced oil-attracting characteristics coupled with water-repelling properties. They employ a combination of two processes—adsorption, absorption—or both, to effectively absorb oil [24]. In the absorption mechanism, oil permeates the pore spaces of the sorbent through intermolecular penetration. The sorbent's capacity to absorb oil is chiefly determined by its porosity and is influenced by specific oil attributes, including viscosity, adhesive properties, and stickiness [8]. Absorption operates on the principle of capillary attraction, with oil infiltrating the material's pores and moving upward (uptake) due to capillary forces. This process can be augmented by pressure and gravitational forces [24]. On the flip side, adsorption involves oil adhering to the surface of the sorbent material, with the extent of adhesion contingent on the viscosity of the oil. Oils with higher viscosity tend to create thicker films on the given material [24]. The bonding takes place due to a range of physical and chemical interactions, such as Van der Waals forces, hydrogen bonding, polarity, steric interactions, hydrophobicity, and others [25].

The sorption process encompasses three distinct phases: (a) oil diffusion onto the sorbent surface, (b) oil entrapment within the sorbent's structure through capillary action, and (c) oil agglomeration within the rough and porous framework of the sorbent [26]. It is understood that both absorption and adsorption occur when oil encounters biopolymeric sorbents. Oil swiftly accumulates on the sorbent surface through lipophilic interactions or coalescence, while absorption transpires as oil migrates into the sorbent material's pores through capillary action [27, 28].

# 3 Types of Biomass Sorbents

Undoubtedly, sorption stands as an effective mechanism for synthetic and inorganic mineral sorbents, given their ability to absorb oil in quantities far exceeding their own weight, thereby transforming it into a semi-solid or solid form suitable for disposal or reuse. Various studies have delved into the sorption capabilities of oleophilic-hydrophobic synthetic materials. However, the substantial chemical alterations these materials undergo have rendered them both ecologically and economically unsustainable. Certain substances within this category can produce toxic byproducts or low molecular weight chemical compounds that pose a threat to aquatic organisms [29]. Moreover, a notable limitation is that many mineral sorbents tend to sink in water and contaminate the ocean floor, compounding the issue. They also exhibit a tendency to release oil during precipitation due to their poor oil retention capacity [29].

There exists a growing interest in identifying cost-effective, abundant, and efficient materials for oil spill sorbents, with a specific emphasis on bio-derived organic sorbents, often derived from agricultural sources. A myriad of keratin and cellulosic materials, including corn stalks [30], cotton [31], milkweed fiber [32], currant fibers [33], kapok [34], banana fibers [35], luffa [36], nettle fibers [37], pistia leaves and roots [38], orange peel [39], palm fibers [40], pineapple leaves [41], pomelo peel [42], rice husks [43], sawdust [44], wheat straw, sugarcane bagasse [45], and walnut shells [28], alongside unconventional sources like human hair, chicken feathers, and wood fibers [46], have exhibited promising outcomes, rendering them viable alternatives for oil spill clean-up.

## *3.1 Cellulose-Based Sorbents*

Cellulose, the main building block found in plant and bacterial cell walls, is the most abundant natural polymer globally, serving as a crucial resource [22]. Cellulosic materials include cellulose itself, along with hemicelluloses and lignin. While lignin repels water, cellulose and hemicellulose tend to attract it, despite being only slightly soluble in water and partially soluble in organic solvents. However, the naturally low hydrophobicity and buoyancy of cellulosic materials limit their effectiveness in separating oil and water. Consequently, researchers are exploring various physical, chemical, and thermal modifications to enhance the hydrophobic and oil-attracting properties of cellulose-based polymers.

Acetylation stands out as a significant method of chemical modification, involving the substitution of functional groups within cellulose and hemicelluloses with hydrophobic counterparts through both catalytic and non-catalytic processes. In acetylation, the acetyl group from acetic anhydride bonds with the hydroxyl group present in cellulose and hemicellulose, resulting in the formation of oleophilic acetate groups ($O–CO–CH_3$). Numerous studies have demonstrated the effectiveness of acetylation in enhancing the hydrophobic nature of cellulosic fibers [35, 47–50]. This chemical alteration involves replacing hydroxyl groups with C=O and C–O groups, thereby increasing the fibers' hydrophobicity and their ability to absorb oil [51, 52]. Additionally, other techniques for chemical modification, such as acid treatment, surfactant treatment, and the deposition of metal oxide nanoparticles, have been utilized [53]. Thermal treatments have also been employed to improve the sorption capacities of cellulosic fibers. These treatments typically decrease the presence of oxygen-containing functional groups and volatile constituents while increasing the carbon content of the cellulose material, thereby enhancing its hydrophobicity [54].

Subsequent sections will delve into diverse cellulose-based biopolymers featured in literature for oil spill sorption.

### 3.1.1 Kapok Fiber

The kapok tree, often referred to as the "Poor man's Silk," produces kapok fiber, consisting mainly of cellulose (64%) and lignin (13%) [55]. Kapok fibers boast distinctive attributes such as low density, sizable lumens, and surfaces coated with wax, all contributing to their exceptional buoyancy, hydrophobicity, and oleophilicity. These qualities enable kapok fibers to repel water and remain afloat after oil recovery [56, 57]. Numerous investigations have delved into the oil sorption capabilities of kapok, consistently revealing its impressive capacity for oil absorption [58, 59]. Considered among the finest biopolymers derived from agricultural waste, kapok stands out due to its robust buoyancy and effective oil sorption performance [57, 60]. The sorption capacity of kapok varies in accordance with the physical and chemical attributes of the absorbed oil. For instance, one study noted a sorption capacity of 50.8 g/g for engine oil on low-density kapok fiber at a low contact angle, while diesel oil exhibited a sorption capacity of 52.7 g/g at a contact angle of 44° [56, 60]. Another study discovered that stirring augmented the adsorption of palm oil onto kapok fiber, with each gram capable of absorbing over 200 times its weight in oil [61]. In terms of modification, researchers have explored the utilization of silica nanoparticles and hydrolyzed dodecyltrimethoxysilane to enhance the sorption capacity of kapok fiber. The modified kapok fiber displayed remarkable oil sorption capability, magnetic responsiveness, and a water contact angle of 143°. Furthermore, it showcased excellent reusability, retaining over 95% of its capacity after 16 cycles of reuse [57].

The elevated sorption capacity of kapok fiber can be attributed to its notable porosity, low cellulose content, broad bamboo-like lumens, and hydrophobicity. Oils adhere to the waxy surfaces of kapok fiber through a combination of hydrophobic interactions and physical forces. Additionally, absorption transpires via the capillary migration of oils into the lumens [62]. The packing density of fiber assemblies influences the sorption capacity, with denser packing resulting in reduced sorption capabilities [63, 64]. Furthermore, the physical attributes of oils, such as viscosity, contact angle, and surface tension, play a role in determining sorption capacities.

Kapok fiber exhibits considerable potential for recovery and reuse, demonstrating successful performance even after 4 to 15 cycles of use [56, 59, 63, 65].

### 3.1.2 Populous Seed Fibre

Populus, also known as cottonwood or poplar tree, is a deciduous tree species that typically attains heights of 20–30 m. It is a prevalent angiosperm within the Salicaceae family and is predominantly found in the northern hemisphere. The tree produces fibrous strands of white hue that harbor micropores. Chemical analysis of these fibers has disclosed their primary composition as lignocellulose. Lignocellulose stands as a composite material encompassing cellulose, hemicellulose, and lignin, collectively responsible for endowing structural integrity to the plant cell wall [66].

Research has unveiled that the fibers of Populus assume the form of hollow hydrophobic microtubes. These hollow fibers exhibit a solid skeleton comprised of lignocellulosic material, while being enveloped by a hydrophobic waxy layer. Diverse studies have reported the remarkable oil sorption attributes of Populus fibers. For example, at a packing density of 0.005 $g/cm^2$, one investigation observed peak sorption capacities of 182–211 g/g and >100 g/g for heavy motor oil and diesel, respectively. The sorption capacity exhibited variation in accordance with the packing density of the fibers [66].

The exceptional oil sorbent properties of Populus fibers find their roots in a combination of chemical, physical, and microstructural characteristics. The presence of microtubes within the fibers permits the entrapping of oil within their hollow lumens, thus contributing to the immensely high sorption capacities [66]. Furthermore, the interplay of cellulose and lignin composition in these hollow fibers is likely instrumental in shaping their sorption prowess. Additionally, Populus fibers showcase notable water resistance and impressive buoyancy, evidenced by a water adsorbency of 3.48% and buoyancy of 90% [25].

### 3.1.3 Cotton Fibre

Cotton, a natural material composed mainly of cellulose (91%), encompasses other constituents including water (7.85%), protoplasm pectin (0.55%), waxes, fatty compounds (0.4%), and mineral salts (0.2%). The sorption prowess of cotton can be ascribed to its substantial cellulose content, which constitutes the predominant portion of the fiber and presents a thicker lumen in the secondary wall. The wax layer enveloping the fiber's primary walls, composed of long-chain saturated fatty acids, alcohols, resins, and sterols, contributes to the fiber's hydrophobic nature, allowing it to attract oil through a non-polar process [67, 68].

Research evaluating raw and acetylated cotton fibers for oil removal from water unveiled that raw cotton exhibited sorption capacities ranging from 19 to 23 g/g, while acetylated cotton displayed capacities varying from 23 to 30 g/g. The fibers could be easily retrieved via simple squeezing and reused for up to ten cycles, with minor reductions in sorption capacities observed after multiple uses. Interestingly, treated cotton exhibited lower sorption capacities than raw cotton fibers, yet it showcased enhanced lipophilic traits and augmented stability in aqueous environments [68].

Subsequent studies have underscored the potential of low-grade cotton fibers and pads for oil spill cleanup, evidencing sorption capacities in the range of 18.43–22.5 g/g. Aligning cotton fibers in a single direction led to a 65% increase in sorption capacity, attributed to the low bulk density of the aligned fibers. Furthermore, cotton fibers have been harnessed to yield carbon fibers with a hollow tubular structure, resulting in sorption capacities spanning 32–77 g/g for diverse oil types. Cotton-derived carbon fibers exhibited heightened surface area and hydrophobicity, proving effective for multiple cycles of oil sorption and recovery [69].

Researchers have also pioneered the creation of superhydrophobic and superoleophilic cotton fibers for oil–water separation, demonstrating water contact angles surpassing 150°, oil contact angles below 5°, and sorption capacities up to 50 g/g [70, 71].

While the oil sorption onto cotton fibers primarily transpires at the fiber surfaces, the significant sorption capacities reported across various studies can be attributed to supplementary processes. These encompass interactions between oils and the wax found in the cotton fiber's main walls, uptake through interfiber capillaries within fiber bundles, and storage within the internal lumens of the fibers. The sorption process is influenced by factors such as the molecular structure and surface morphology of the fibers, as well as the characteristics of the oil being sorbed [67].

### 3.1.4 Cattail Pollen Fibre

Cattail fruit produces cattail fiber, a lignocellulosic material containing 10.64% wax content. The fiber is characterized by brittleness and shortness, and it is primarily utilized as filler for heat retention and life preservers [34, 56]. Given its remarkable buoyancy and hydrophobic nature, its potential for oil–water separation has garnered attention [34][34].

In a study conducted by Khan et al. (2004), the oil sorption and retention capabilities of various biomass sorbents, including commercial polyester fiber, were examined using engine oil as the sorbate. The sorption capacity of cattail fiber was measured at 11.07 g/g, slightly surpassing the sorption capacity of polyester fiber at 10.08 g/g. In another experiment, cattail fiber displayed sorption capacities of 13.4 g/g for engine oil and 14.6 g/g for vegetable oil, accompanied by retention capacities exceeding 95% after a 24-h drainage period [72, 73].

The noteworthy high sorption capacities of cattail fiber have been attributed to its elevated surface waxiness, hydrophobicity, and oleophilic traits [74]. Cao et al. (2016–2017) also delved into the oil sorption attributes of cattail fiber, revealing a sorption capacity of 12 g/g for vegetable oil after 24 h, coupled with over 88% retention. Cattail fiber exhibited hydrophobic characteristics comparable to kapok, characterized by a water contact angle of 132.78°. The robust oleophilic and water-repellent attributes of the fiber's surface were attributed to its substantial hydrophobic wax content. The bamboo-shaped structure of cattail fiber offered extensive exposed areas, further enhancing its capacity for oil sorption [31].

### 3.1.5 Rice Husk/Straw

Rice husk, a byproduct of rice milling in agriculture, is composed of lignin (28.25%), hemicellulose (18.59%), and cellulose (31.13%), with a bulk density ranging from 34 to 40 kg/m$^3$ [50].

Sun et al. (2002) conducted a study on the adsorption of machine oil using acetylated rice straw with and without a catalyst. Acetylation without a catalyst led to a weight increase of 11.2%, and the sorption capacity ranged from 16.8 to 20 g/g. The sorption capabilities of acetylated rice husk were influenced by the degree of acetylation [75].

Rice husk also contains a notable amount of silica. The pyrolysis of rice husk results in the breakdown of organic matter and the release of silicon-organic bonds, enhancing the hydrophobic nature of rice husk and its potential for crude oil purification [76]. The adsorption of oils is contingent upon factors such as granulometric composition (C/$SiO_2$ weight ratio), sorbent bulk density, surface morphology, and oleophilicity. An inverse relationship exists between sorbent bulk density and oil penetration depth [77]. Alkaline treatment of rice husk eliminates silica and improves the absorption capacity for diesel, rendering it suitable as a solid fuel in conjunction with the sorbent [78].

Uzunov et al. (2012) examined the impact of slow pyrolysis temperatures (250–700 °C) on the crude oil sorption capacities of rice husk. At 350 °C, pyrolyzed rice husk exhibited a sorption capacity of approximately 10 g/g, in contrast to 6.5 g/g for raw rice husk. The researchers concluded that the porous structure of pyrolyzed rice husk played a more pivotal role in oil absorption capabilities than the quantity of surface functional groups [79].

### 3.1.6 Bacteria Cellulose

Bacterial or microbial cellulose (BC) can be synthesized by bacteria like Acetobacter, Sarcinaventriculi, and Agrobacterium. BC stands out for its purity, lacking the lignin and other impurities typically found in plant cellulose. It boasts a distinct nanostructure comprised of interconnected cellulose nanofibers [80]. BC showcases an ultrafine network architecture, remarkable mechanical strength, and notable water holding capacity [81, 82].

Sai et al. (2015) conducted modifications on the surface of BC aerogel using trimethylchlorosilane, yielding exceptional superhydrophobic and oleophilic attributes, marked by a water contact angle of 146.5° [83]. The BC aerogel displayed sorption capacities spanning from 86 to 185 g/g for diverse oils, showcasing commendable recyclability. Even after undergoing ten cycles of reuse, no significant decline in sorption performance was observed.

## *3.2 Keratin-Based Sorbents*

Keratins are structural proteins that create networks linked by disulfide bonds. Alongside collagen, they rank among the most crucial animal biopolymers. Keratins manifest in diverse forms like feathers, quills, wool, claws, beaks, and hair. These substances consist of intermediate filaments crystalline on the nanoscale to centimeter scale, enveloped by an amorphous protein matrix [84].

The potential of keratin-based biopolymers for oil–water separation processes has been explored [85–87]. Different keratinous materials share analogous primary and secondary protein structures, characterized by a polypeptide backbone arising from the condensation of L-amino acids. These biopolymers can sorb pollutants, including oil-based contaminants, through physiosorption and chemisorption mechanisms. Sorption transpires through the material's porous network, its surface, or chemical interactions with functional groups like peptides and amino acids inherent in keratin proteins [84, 87]. Hair, wool, and chicken feathers have undergone extensive study as potential oil sorbents. These materials can be employed in their natural states or subjected to diverse treatments to heighten their oleophilicity and hydrophobicity, akin to the treatments discussed in Sect. 3 for cellulose-based materials.

### 3.2.1 Human Hair

Human hair, a readily available natural waste material, comprises keratin fibers containing protein, cuticles, water, lipids, and trace elements. The porosity of hair protein is determined by the end bonds of amino acids. Hair keratin is acknowledged for its hardness, water resistance, and durability, constituting an insoluble protein composed of 18 amino acids. The hydrophobic character of hair is owed to its cuticle and lipids. Human hair waste possesses distinctive surface tensile strength and recovery elastic limit, enabling it to repel water while absorbing oils, rendering it suitable for oil adsorption [87, 88].

The amalgamation of keratin and lipid-guarded cuticles in human hair underscores its high oil absorption and low water absorption capacities. The hydrophobic traits stem from the presence of medulla, cortex, and dead cells in the cuticle, preventing water absorption [89]. Jadhav et al. scrutinized the oil removal efficiency of black, brown, and blonde hair in an aqueous milieu [88]. They discerned that black hair showcased the highest adsorption capacity, suggesting stronger adhesive forces between oil and hair than between water and hair. Ingoles et al. reported hair oil sorption from wastewater, noting reduced sorption capacity with elevated pH levels. Men's hair displayed a higher absorption capacity (0.014 g/g) compared to women's hair (0.010 g/g).

### 3.2.2 Chicken and Avian Feathers

The chicken food industry generates a substantial volume of chicken feathers (CFs) as waste by-products, with an annual output of 5 million tonnes. CFs are composed of 91% keratin, existing in alpha-helix, beta-sheet, and disordered forms [90]. The chemical composition of CFs includes 2.9% sulfur, 15.5% nitrogen, 15.8% lipids, and 1% ash [91]. Owing to their composition, CFs have been explored as potential low-cost sorbents for wastewater treatment [90, 91] and oil sorption applications [92].

CFs feature knotted and hooked structures with hollow spaces, contributing to an amplified pore volume available for oil adsorption [92]. The interconnected pore structure and rough surface of CFs bolster oil retention within the sorbent material. Chinonyere examined the oil sorption on CFs within an oil/water mixture and identified the highest sorption capacities as 7.7 g/g for vegetable oil, 6.1 g/g for crude oil, and 4.1 g/g for diesel fuel [92]. The heightened adsorption capacity of vegetable oil and crude oil can be ascribed to their elevated viscosities, promoting more robust interactions with the active sites on CFs. The adsorption kinetics exhibited an inverse correlation with increasing temperature and were best represented by a pseudo-second-order kinetic model.

Wang et al. harnessed graft copolymerization to extract keratin from chicken feathers, subsequently treating it with oleic acid to fabricate a keratinous cryogel [93]. The cryogel was then enhanced via glutaraldehyde crosslinking for motor oil sorption, resulting in a maximum sorption capacity of 10.76 g/g. The enhanced sorption capacity can be attributed to the crosslinking of glutaraldehyde with hydrophobic groups and the formation of fibrous structures within the protein matrix.

### 3.2.3 Wool Fibre

Wool, similar to human hair, is composed of keratin with a comparable chemical composition. However, the physical structure of wool sets it apart, characterized by a higher number of scaled cells and an indented free edge. This unique structure allows wool fibers to interlock, resulting in a highly porous and swollen state that enhances water sorption. Wool has been noted for its affinity for oil, with its oil sorption efficiency attributed to factors such as surface roughness, porosity, and the presence of natural substances like lanolin and waxes on the fiber surface [85].

Rajakovic-Ognjanovic delved into the factors influencing the removal of machine oil from water using natural wool fiber (NWF) and recycled wool-based non-woven material (RWNM) [86]. The sorption capacities of NWF and RWNM fell within the ranges of 33–43 g/g and 15–19 g/g, respectively. These capacities were influenced by pH and temperature, and the sorption process was determined as physisorption. Treatments like chitosan and low-temperature air plasma applied to RWNM did not enhance oil adsorption, as these treatments altered the lanolin and waxes on the fiber surfaces, which were partially responsible for the oil sorption performance.

Collectively, wool exhibits favorable traits for oil sorption, including its surface roughness, porosity, and the presence of natural compounds like greases and waxes. These characteristics contribute to its effective oil sorption performance, positioning it as a promising material for oil spill cleanup and various oil–water separation applications.

## 4 Factors Affecting Oil Sorption

The substrate properties of sorbent materials are considered the most important factor in determining the quantity of oil absorbed and retained in the porous structure of the absorbent. Factors such as porosity, surface area, hydrophobicity, and surface roughness contribute to the oil absorption capacity of the material. The characteristics of the spilled oil also play a significant role in determining the effectiveness of oil cleanup. Properties such as viscosity (thickness), density, volatility (evaporation rate), and chemical composition affect the interaction between the oil and the absorbent material, influencing the absorption and retention capabilities. The marine environment, including factors like wind, waves, and temperature, can impact the efficiency of oil spill cleanup. These environmental conditions affect the movement and spreading of the spilled oil, as well as the behavior of the absorbent material in capturing and containing the oil. Table 1 provides a complete analysis of the variables influencing the oil absorption capacity of absorbents made of natural materials.

Overall, the combination of substrate properties, spilled oil characteristics, and the marine environment determines the effectiveness and efficiency of natural absorbents in oil absorption and cleanup operations.

## 5 Modification Methods for Hydrophobicity Improvement

While natural-based materials are abundant and environmentally friendly, they often exhibit lower sorption performance in comparison to synthetic counterparts due to their inherent deficiencies in oleophilic and hydrophobic characteristics. To address these limitations and enhance their efficacy in oil sorption applications, a range of modification techniques can be utilized. Mechanical methods, like grinding or crushing, can augment sorbent adsorption capacity by increasing surface area, though they have limited impact on material hydrophobicity. Thermal and hydrothermal modification methods involve subjecting the sorbent material to elevated temperatures or steam treatment, respectively. These procedures can induce structural changes that lead to improved sorption properties, encompassing heightened hydrophobicity and expanded porosity. Chemical modification techniques offer a versatile toolkit for enhancing the sorption abilities of natural-based materials. Examples include processes such as mercerization, acetylation, benzoylation, and grafting. These techniques involve treating the material with specific chemicals or

**Table 1** Factors affecting the absorption capacity of natural absorbents [94–97]

| Properties | Explanations |
|---|---|
| *Properties of absorbent* | |
| Types of absorbent | • Chemically treated hydrophobic cellulosic fibre products have greater absorption capacity than biomass materials<br>• Other factors influencing absorption capacity include size, wetness, density, and chemical composition |
| Surface area | • The absorption capacity increases (when a unit mass is taken into account) with increasing accessible surface area<br>• The more viscous oil spill can be absorbed by the larger pores |
| Surface area | • The absorption capacity increases (when a unit mass is taken into account) with increasing accessible surface area<br>• The more viscous oil spill can be absorbed by the larger pores |
| Drying process | • By using the standard drying method, the porous structure of natural sorbent materials can be significantly decreased, reducing their absorption capacity and cycle of use |
| Hold up and release of the absorbed oil | • The pore-like structure of a cellulose-based absorbent material is significant in relation to the observed absorption<br>• The main components in absorbing and storing the absorbed oil can alternatively be cellulose-based absorbent materials having a mesoporous structure<br>• In contrast, the relaxation of tensions brought on by the biological decomposing of natural absorbent materials results in the release of absorbed oil |
| The lumen of the fibre | • In comparison to the measured absorption, there is a large amount of oil absorption into the lumens of cellulose-based absorbent materials |
| SpringBack | • It is described as the capacity of a porous volume to recover following compression. The elastic recovery of absorbent materials made of cellulose is significant |
| Floating ability | • An absorbent's ability to float on saltwater is both a desired and necessary quality. If the absorbent sinks, the absorbed oil will be drawn into its pores and damage the environment |
| Bulk density and available pore volume | • Sorbent materials that resist tight packing have a better oil absorption capability<br>• Bigger pores encourage both a greater rate of fluid oil evaporation and a higher absorption capacity<br>• The cross-linking, which is limiting or lowering the swell of the absorbent, also lessens the pickup of spilt oil |
| *Properties of spilled oil* | |
| Triple physical property (viscosity, density, surface tension) | • Low-triple physical features of the oil show a more quick absorption. In some circumstances, absorbent materials with wide pores are thought to be acceptable for the oil's higher-than-triple physical qualities. In general, the three physical characteristics of oil spills have a significant impact on how well absorbent materials can absorb them |

(continued)

**Table 1** (continued)

| Properties | Explanations |
|---|---|
| State of oxidation/ evaporation, emulsification | • Recovery of spilt oil that takes a long period makes the oil more viscous, which leads to oil oxidation, evaporation, emulsification, and loss of absorbed oil |
| The thickness of the oil spill | • Accurately measuring the thickness of the spilled oil is crucial for improving the efficiency of recovered oil |
| Water take | • Oil absorption usually happens concurrently with water uptake |
| *Properties of marine environment* | |
| pH, temperature, salinity, wind velocity, and wave height | • Wave height, wind speed, pH, temperature, and salinity all have a significant impact on the capacity to absorb oil. Strong waves, high winds, and high temperatures all lower the oil absorption capability. Natural absorbent materials can be used when the pH is between 4 and 6. Salinity in saltwater causes an increase in water intake, which reduces oil absorption |

reactants to alter its chemical structure, surface characteristics, and hydrophobic properties, thereby bolstering its sorption performance. By deploying these modification strategies, natural-based sorbents can be customized to exhibit enhanced sorption capacities, elevated hydrophobicity, and superior performance in oil sorption applications. Such modifications enable the effective utilization of abundant and renewable natural resources, facilitating efficient oil spill cleanup and environmental remediation.

## 5.1 Physical and Thermal Treatment

Grinding natural sorbent media has been proven to enhance their oil sorption capacity by increasing the accessibility of the contact surface and providing additional binding sites on smaller particles. Bayat et al. conducted experiments involving powdered bagasse, a byproduct of sugarcane production, with varying particle sizes to assess the impact of particle size on oil spill cleanup [98]. The findings indicated that the effectiveness of oil separation improved as the fibrous bio-sorbents took the form of fine fibers.

Low-temperature thermal treatment and drying have a limited effect on the oleophilic and hydrophobic characteristics of the sorbent, primarily refining the surface contaminants and facilitating smoother oil absorption. Conversely, high-temperature thermal treatment, such as pyrolysis [99], induces the carbonization of the sorbent, substantially augmenting its oil sorption capacity and oil-to-water selectivity [77].

In Husseien et al.'s study, the sorption properties of pyrolyzed barley straw were examined at different temperatures [100]. The oil sorption capacity measured 7.6 g/

g and 9.2 g/g for diesel and heavy oil, respectively. Carbonized barley straw exhibited pronouncedly hydrophobic characteristics and absorbed larger quantities of oil compared to commercial polypropylene pads. Additionally, the carbonized barley straw proved reusable across multiple absorption and desorption cycles.

Surface alterations resulting from thermally treated natural-based materials can be observed through scanning electron microscopy (SEM) images. El Gheriany's research revealed that thermal decomposition of orange peel at 500 °C yielded a rough and notably porous structure as opposed to its original smooth and homogeneous state [39]. The creation of pores is attributed to material decomposition and the release of gases/vapors during pyrolysis.

These investigations underscore the potential of thermal treatment and pyrolysis to modify natural-based sorbents, heightening their oil sorption capacity and hydrophobicity. The resultant materials hold promise for oil spill cleanup applications and offer reusability across multiple cycles of oil absorption and desorption.

## 5.2 Hydrothermal Treatment

Hydrothermal treatment presents an alternative avenue to thermal pyrolysis in the creation of carbonaceous compounds, particularly concerning natural sorbents derived from biomass. This method involves subjecting the material to subcritical or supercritical water conditions. Research has illustrated that hydrothermal treatment diminishes oxygen-containing functional groups and volatile components within the biomass. This transformative process augments the carbon content and heightens the hydrophobic attributes of the resultant sorbent materials. A notable benefit of hydrothermal treatment is its capacity to obviate the need for preliminary biomass drying, which can be laborious and energy-intensive. Biomass materials such as wood, plants, and vegetables can possess substantial water content (up to 70% by weight). Conventional thermal processes necessitate prior drying of the material. Yet, hydrothermal treatment engages the biomass within a solution, bypassing the drying phase and curbing overall processing time and energy demands [101–103].

Li et al. conducted an investigation into the efficacy of winter melon aerogel generated through hydrothermal treatment for oil sorption [71]. The resultant aerogel exhibited a density of 0.048 g/cm$^3$ and a remarkable absorption capacity of 25 g/g. Aerogels originating from natural sources display pronounced hydrophobic characteristics with a contact angle of 135°. Coupled with low density and high porosity, these attributes contribute to their outstanding oleophilic properties. Hydrothermal treatment presents an auspicious avenue for producing carbonaceous sorbents from biomass, conferring heightened hydrophobicity and oil sorption capacity. It emerges as a more efficacious and cost-effective alternative to traditional thermal processes, particularly when dealing with biomass materials characterized by elevated water content [104].

## 5.3 Chemical Modification Methods

### 5.3.1 Mercerization

Cellulose, a polymer comprised of β-glucose units linked by β-1,4 glycosidic bonds, assumes various crystalline forms, with Cellulose I and II being of primary significance [105]. The innate hydrophilic nature of cellulose poses challenges for its efficacy in separating oil and water, necessitating the infusion of hydrophobic and oleophilic attributes. While Cellulose I is naturally prevalent in fibers, its crystal arrangement can be altered through procedures like mercerization [106, 107]. This entails subjecting fibers to a hot or cold NaOH solution, thereby enhancing impurity removal and transforming the crystalline structure from Cellulose I to Cellulose II. This shift results in a more textured and rigid surface, bolstering contact area and augmenting adsorption efficiency.

In the study conducted by Gulati et al., the influence of mercerization on sawdust was scrutinized, revealing pronounced morphological transformations via SEM analysis [108]. This surface alteration facilitates improved interaction between the material and oil, endowing mercerized cellulose with advantages for aqueous oil spill cleanup. The technique has also been successfully applied to cattail fibers and soybean waste, heightening their sorption capacities for motor oil and crude oil, respectively.

Rice husks have found utility as feedstock for producing mercerized natural-based oil sorbents. Alkali treatment induces structural modifications in rice husk, yielding a material endowed with appreciable oil sorption capability. The process operates at modest temperatures, ensuring favorable product yields [78]. BET and FTIR analyses underscore that microporosity and surface functional groups are not the principal determinants of oil sorption capacity. Notably, the sorption capacity for marine diesel exhibits an inverse correlation with bulk density. Mercerized rice husk, characterized by lower bulk density, fosters internal oil diffusion within the substrate.

### 5.3.2 Acetylation

Acetylation stands as a method utilized to bolster the sorption efficacy of sorbents, achievable with or without a catalyst [109]. This process entails transforming hydroxyl groups within the cellulose structure into oleophilic acetate groups (O–CO–$CH_3$) via a reaction with acetic anhydride. This metamorphosis curtails the hydrophilic attributes of the sorbent while amplifying its oleophilic traits [110]. Numerous investigations have attested to the potency of acetylation in augmenting the hydrophobicity of natural fibers. To illustrate, acetylated rice straw underwent scrutiny in a study that encompassed both catalytic and non-catalytic methodologies. Devoid of a catalyst, acetylation yielded an 11.2% uptick in sorbent mass and yielded oil sorption efficiencies spanning 16.8–20 g/g. In the catalyst-aided scenario (utilizing 4-dimethylaminopyridine), acetylation yielded a 15.4% mass increase and oil sorption capacities ranging from 20.9 to 24 g/g. Notably, the extent of acetylation exhibited a positive correlation with sorption efficiency for rice husk [111].

The scope of acetylation expanded to augment the hydrophobicity of kapok fibers, acknowledged for their superior buoyancy and substantial oil absorption capabilities. Wang et al. achieved successful acetylation of kapok fibers, with the modified fibers showcasing elevated diesel oil absorption capacity (36.7 g/g) relative to unaltered fibers (27 g/g). The integration of nonpolar acetyl groups on the surface of modified kapok fibers endowed them with hydrophobicity, thereby facilitating favorable oil absorption. SEM images that juxtaposed raw and acetylated kapok fibers corroborated these findings, revealing a more uneven surface and open lumen orifice in the modified fibers—features conducive to oil sorption [57]. Acetylation has traversed the realms of modifying other materials, including cotton fibers and corn straw fibers, engendering heightened efficiency and bolstered sorption capacity.

### 5.3.3 Grafting

Polymerization emerges as a widely employed strategy for grafting monomers onto the exteriors of fibers, whether they are natural or synthetic. This technique finds application across diverse sorbents, encompassing polypropylene, banana fibers, and coir. Noteworthy achievements in oil sorption capacities have been attained via grafting, notably reaching 14.45 g/g for grafted banana fibers [35] and 13.45 g/g for grafted coir [48]. Within the realm of banana fibers, grafting entails the establishment of esters that supplant the hydroxyl groups present on the cellulose surface [35]. Maintaining a judicious proportion of the synthetic polymer chain in the grafted material is of paramount significance, as this balance ensures the absence of interference with the expansion of separated fibers in aqueous environments. This meticulous control holds considerable importance in the effective extraction of oil from water, facilitated by the grafted natural-based sorbent.

Employing Scanning Electron Microscopy (SEM), scrutiny of both unmodified and grafted fibers' surface structures has been undertaken. SEM images lay bare pronounced distinctions in the surfaces of these two fiber categories. Chemically grafted banana fibers exhibit a heterogeneous overlay of poly butyl acrylate species on their surface, with certain fibers potentially displaying cross-linking due to the grafting reaction. This alteration imparts bulkiness and heightened hydrophobicity to the fibers, resulting in improved oil absorption attributes. Concordant outcomes have surfaced across other natural-based sorbents subjected to grafting methodologies.

## 6 Conclusions and Future Aspect

The primary objective of this study is to offer a comprehensive overview of the cutting-edge biosorption technology landscape, encompassing a broad spectrum of scholarly articles and patents linked to the subject matter. Biosorption has garnered considerable attention across academic, research, and industrial domains due to its expanding utility across diverse sectors. The pursuit of utilizing cost-efficient absorbents characterized by robust oil absorption capabilities for oil spill recovery stands as a highly desirable aspiration, given its alignment with requisites such as cost-effectiveness, minimal ecological impact, and effective waste handling.

Central to this chapter's focus are non-traditional, cost-effective adsorbents centered on cellulose and keratin, spotlighting their applications in oil spill cleanup and oil–water separation methodologies. The study yields several key findings:

1. Non-traditional adsorbents boasting attributes like biodegradability, natural abundance, affordability, and environmental compatibility outperform conventional counterparts when it comes to oil spill recuperation.
2. The ideal oil spill sorbent should exhibit tailored physical attributes, including lipophilicity, hydrophobicity, biodegradability, and a substantial capacity for absorption and retention.
3. Techniques like mechanical, thermal, and chemical modification can effectively enhance the subpar oleophilic and hydrophobic traits inherent to sorbents.
4. The sorbent material's properties, the characteristics of the spilled oil, and the marine environment's conditions collectively shape the oil absorption capacity.
5. Integral to the broader context, economic analyses hold significance in comprehending the comprehensive cost associated with both the sorbent and biosorption process, particularly when scaling up biosorption technology for industrial wastewater treatment.
6. Upcoming research endeavors should emphasize chemical refinements of biosorbents to amplify their oil absorption prowess. Exploring advanced technologies is paramount to refining bioadsorbents' capability to absorb oil spills while minimizing adverse repercussions on marine ecosystems and human well-being.

In summation, this study accentuates the substantial promise inherent in biosorption technology, underscoring the ongoing necessity for continuous research and development to optimize its efficacy and pragmatic application across real-world scenarios.

## References

1. [Internet]. Available from: https://response.restoration.noaa.gov/training-and-education/education-students-and-teachers/how-do-spills-happen.html

2. Reddy CM, Eglinton TI, Hounshell A, White HK, Xu L, Gaines RB et al (2002) The West Falmouth oil spill after thirty years: the persistence of petroleum hydrocarbons in marsh sediments. Environ Sci Technol 36:4754–4760
3. Liu Z, Liu J, Zhu Q, Wu W (2012) The weathering of oil after the Deepwater Horizon oil spill: insights from the chemical composition of the oil from the sea surface, salt marshes and sediments. Environ Res Lett 7:035302
4. Doshi B, Sillanpää M, Kalliola S (2018) A review of bio-based materials for oil spill treatment. Water Res 135:262–277
5. Sarbatly R, Krishnaiah D, Kamin Z (2016) A review of polymer nanofibres by electrospinning and their application in oil–water separation for cleaning up marine oil spills. Mar Pollut Bull 106:8–16
6. Barron MG (2012) Ecological impacts of the Deepwater Horizon oil spill: implications for immunotoxicity. Toxicol Pathol 40:315–320
7. Brody TM, Di BP, Krysa J (2012) Analysis of inland crude oil spill threats, vulnerabilities, and emergency response in the Midwest United States. Risk Anal 32:1741–1749
8. Ifelebuegu AO, Johnson A (2017) Nonconventional low-cost cellulose- and keratin-based biopolymeric sorbents for oil/water separation and spill cleanup: a review. Crit Rev Environ Sci Technol 47:964–1001
9. Eze SI, Akpomie KG, Ezeofor CC, Mmadubuike NV, Ojo FK (2019) Remediation of oil spill polluted water from Niger Delta Nigeria by sorption onto ammonium sulfate modified *Dialium guineense* seed husk. Pet Sci Technol 37:1830–1838
10. Motta FL, Stoyanov SR, Soares JBP (2019) Development and application of an amylopectin-graft-poly(methyl acrylate) solidifier for rapid and efficient containment and recovery of heavy oil spills in aqueous environments. Chemosphere 236:124352
11. Hoang AT, Le VV, Al-Tawaha ARMS, Nguyen DN, Al-Tawaha ARMS, Noor MM et al (2018) An absorption capacity investigation of new absorbent based on polyurethane foams and rice straw for oil spill cleanup. Pet Sci Technol 36:361–370
12. (2020) The use of dispersants in marine oil spill response. National Academies Press, Washington, D.C.
13. Ogbonna D (2018) Application of biological methods in the remediation of oil polluted environment in Nigeria. J Adv Biol Biotechnol 17:1–10
14. Sayed K, Baloo L, Sharma NK (2021) Bioremediation of total petroleum hydrocarbons (TPH) by bioaugmentation and biostimulation in water with floating oil spill containment booms as bioreactor basin. Int J Environ Res Public Health 18:2226
15. Hoang AT, Pham XD (2021) An investigation of remediation and recovery of oil spill and toxic heavy metal from maritime pollution by a new absorbent material. J Mar Eng Technol 20:159–169
16. Fingas M (ed) (2016) Oil spill science and technology, 2nd edn. Gulf Professional Publishing
17. Bidgoli H, Khodadadi AA, Mortazavi Y (2019) A hydrophobic/oleophilic chitosan-based sorbent: toward an effective oil spill remediation technology. J Environ Chem Eng 7:103340
18. Tayeb AM, Farouq R, Mohamed OA, Tony MA (2020) Oil spill clean-up using combined sorbents: a comparative investigation and design aspects. Int J Environ Anal Chem 100:311–323
19. Chau MQ, Truong TT, Hoang AT, Le TH (2021) Oil spill cleanup by raw cellulose-based absorbents: a green and sustainable approach. Energy Sourc Part A Recover Util Environ Eff 1–14
20. Karan CP, Rengasamy R, Das D (2011) Oil spill cleanup by structured fibre assembly. Indian J Fibre Text Res 36:199–200
21. Osamor A, Momoh Z (2015) An evaluation of the adsorptive properties of coconut husk for oil spill cleanup. In: International conference on advances in applied science and environmental technology—ASET 2015. Institute of Research Engineers and Doctors, pp 33–37
22. Abdel-Moghny T, Keshawy M (eds) (2014) An overview on the treatment of oil spill using sorbent materials: synthetic and natural oil sorbent. LAP LAMBERT Academic Publishing, Germany

23. Zhu H, Qiu S, Jiang W, Wu D, Zhang C (2011) Evaluation of electrospun polyvinyl chloride/polystyrene fibers as sorbent materials for oil spill cleanup. Environ Sci Technol 45:4527–4531
24. Idris J, Eyu GD, Mansor AM, Ahmad Z, Chukwuekezie CS (2014) A preliminary study of biodegradable waste as sorbent material for oil-spill cleanup. Sci World J 2014:1–5
25. Sabir S (2015) Approach of cost-effective adsorbents for oil removal from oily water. Crit Rev Environ Sci Technol 45:1916–1945
26. Tansel B, Pascual B (2011) Removal of emulsified fuel oils from brackish and pond water by dissolved air flotation with and without polyelectrolyte use: Pilot-scale investigation for estuarine and near shore applications. Chemosphere 85:1182–1186
27. Inagaki M, Kawahara A, Nishi Y, Iwashita N (2002) Heavy oil sorption and recovery by using carbon fiber felts. Carbon N Y 40:1487–1492
28. Srinivasan A, Viraraghavan T (2008) Removal of oil by walnut shell media. Bioresour Technol 99:8217–8220
29. Al-Majed AA, Adebayo AR, Hossain ME (2012) A sustainable approach to controlling oil spills. J Environ Manage 113:213–227
30. Husseien M, Amer AA, El MA, Hamedallah N (2015) Oil spill removal from water by using corn stalk: factors affecting sorption process. Int J Environ Waste Manag 16:281
31. Cao S, Dong T, Xu G, Wang F (2017) Oil spill cleanup by hydrophobic natural fibers. J Nat Fibers 14:727–735
32. Panahi S, Moghaddam MK, Moezzi M (2020) Assessment of milkweed floss as a natural hollow oleophilic fibrous sorbent for oil spill cleanup. J Environ Manage 268:110688
33. Elias E, Costa R, Marques F, Oliveira G, Guo Q, Thomas S et al (2015) Oil-spill cleanup: the influence of acetylated curaua fibers on the oil-removal capability of magnetic composites. J Appl Polym Sci 132
34. Dong T, Xu G, Wang F (2015) Oil spill cleanup by structured natural sorbents made from cattail fibers. Ind Crops Prod 76:25–33
35. Teli MD, Valia SP (2013) Acetylation of banana fibre to improve oil absorbency. Carbohydr Polym 92:328–333
36. Annunciado TR, Sydenstricker THD, Amico SC (2005) Experimental investigation of various vegetable fibers as sorbent materials for oil spills. Mar Pollut Bull 50:1340–1346
37. Viju S, Thilagavathi G, Vignesh B, Brindha R (2019) Oil sorption behavior of acetylated nettle fiber. J Text Inst 110:1415–1423
38. Sánchez-Galván G, Mercado FJ, Olguín EJ (2013) Leaves and roots of *Pistia stratiotes* as sorbent materials for the removal of crude oil from saline solutions. Water, Air, Soil Pollut 224:1421
39. El Gheriany IA, Ahmad El Saqa F, Abd El Razek Amer A, Hussein M (2020) Oil spill sorption capacity of raw and thermally modified orange peel waste. Alexandria Eng J 59:925–932
40. Abdelwahab O, Nasr SM, Thabet WM (2017) Palm fibers and modified palm fibers adsorbents for different oils. Alexandria Eng J 56:749–755
41. Cheu SC, Kong H, Song ST, Saman N, Johari K, Mat H (2016) High removal performance of dissolved oil from aqueous solution by sorption using fatty acid esterified pineapple leaves as novel sorbents. RSC Adv 6:13710–13722
42. Chai W, Liu X, Zou J, Zhang X, Li B, Yin T (2015) Pomelo peel modified with acetic anhydride and styrene as new sorbents for removal of oil pollution. Carbohydr Polym 132:245–251
43. Kenes K, Yerdos O, Zulkhair M, Yerlan D (2012) Study on the effectiveness of thermally treated rice husks for petroleum adsorption. J Non Cryst Solids 358:2964–2969
44. Nurliyana Che Mohamed Hussein S, Hidayati Othman N, Dollah A, Nazihah Che Abdul Rahim A, Shuhadah Japperi N, Syamimi Mohd Asymawi Ramakrishnan N (2019) Study of acid treated mixed sawdust as natural oil sorbent for oil spill. Mater Today Proc 19:1382–1389
45. Behnood R, Anvaripour B, Jaafarzadeh N, Farasati M (2016) Oil spill sorption using raw and acetylated sugarcane bagasse. J Cent South Univ 23:1618–1625
46. Zamparas M, Tzivras D, Dracopoulos V, Ioannides T (2020) Application of sorbents for oil spill cleanup focusing on natural-based modified materials: a review. Molecules 25:4522

47. Sanket P, Valia TM (2013) Application of modified coir fiber as eco-friendly oil sorbent. J Fash Technol Text Eng 01
48. Teli MD, Valia SP, Mifta J (2016) Application of functionalized coir fibre as eco-friendly oil sorbent. J Text Inst 1–6
49. Asadpour R, Sapari N, Hasnain Isa M, Kakooei S, Orji KU, Daneshfozoun S (2016) Esterification of corn silk fiber to improve oil absorbency. Adv Mater Res 1133:552–556
50. Wang Z, Barford JP, Hui CW, McKay G (2015) Kinetic and equilibrium studies of hydrophilic and hydrophobic rice husk cellulosic fibers used as oil spill sorbents. Chem Eng J 281:961–969
51. Banerjee SS, Joshi MV, Jayaram RV (2006) Treatment of oil spill by sorption technique using fatty acid grafted sawdust. Chemosphere 64:1026–1031
52. Said A, Ludwick A, Aglan H (2009) Usefulness of raw bagasse for oil absorption: a comparison of raw and acylated bagasse and their components. Bioresour Technol 100:2219–2222
53. Zang D, Zhang M, Liu F, Wang C (2016) Superhydrophobic/superoleophilic corn straw fibers as effective oil sorbents for the recovery of spilled oil. J Chem Technol Biotechnol 91:2449–2456
54. Liu H, Geng B, Chen Y, Wang H (2017) Review on the aerogel-type oil sorbents derived from nanocellulose. ACS Sustain Chem Eng 5:49–66
55. Reddy N, Yang Y (2009) Properties and potential applications of natural cellulose fibers from the bark of cotton stalks. Bioresour Technol 100:3563–3569
56. Abdullah MA, Rahmah AU, Man Z (2010) Physicochemical and sorption characteristics of Malaysian *Ceiba pentandra* (L.) Gaertn. as a natural oil sorbent. J Hazard Mater 177:683–691
57. Wang J, Zheng Y, Wang A (2012) Superhydrophobic kapok fiber oil-absorbent: preparation and high oil absorbency. Chem Eng J 213:1–7
58. Hori K, Flavier ME, Kuga S, Lam TBT, Iiyama K (2000) Excellent oil absorbent kapok [*Ceiba pentandra* (L.) Gaertn.] fiber: fiber structure, chemical characteristics, and application. J Wood Sci 46:401–404
59. Dong T, Cao S, Xu G (2017) Highly efficient and recyclable depth filtrating system using structured kapok filters for oil removal and recovery from wastewater. J Hazard Mater 321:859–867
60. Rengasamy RS, Das D, Praba KC (2011) Study of oil sorption behavior of filled and structured fiber assemblies made from polypropylene, kapok and milkweed fibers. J Hazard Mater 186:526–532
61. Quek CS, Ngadi N, Zaini MAA, Ramakrishna S (2014) Stirring enhances removal of oil by kapok fiber. Appl Mech Mater 695:69–72
62. Dong T, Wang F, Xu G (2015) Sorption kinetics and mechanism of various oils into kapok assembly. Mar Pollut Bull 91:230–237
63. Lim T-T, Huang X (2007) Evaluation of kapok (*Ceiba pentandra* (L.) Gaertn.) as a natural hollow hydrophobic–oleophilic fibrous sorbent for oil spill cleanup. Chemosphere 66:955–963
64. Dong T, Cao S, Xu G (2016) Highly porous oil sorbent based on hollow fibers as the interceptor for oil on static and running water. J Hazard Mater 305:1–7
65. Wang J, Zheng Y, Wang A (2013) Investigation of acetylated kapok fibers on the sorption of oil in water. J Environ Sci 25:246–253
66. Likon M, Remškar M, Ducman V, Švegl F (2013) Populus seed fibers as a natural source for production of oil super absorbents. J Environ Manage 114:158–167
67. Carmody O, Frost R, Xi Y, Kokot S (2008) Selected adsorbent materials for oil-spill cleanup. J Therm Anal Calorim 91:809–816
68. Deschamps G, Caruel H, Borredon M-E, Bonnin C, Vignoles C (2003) Oil removal from water by selective sorption on hydrophobic cotton fibers. 1. Study of sorption properties and comparison with other cotton fiber-based sorbents. Environ Sci Technol 37:1013–1015
69. Wang B, Karthikeyan R, Lu X-Y, Xuan J, Leung MKH (2013) Hollow carbon fibers derived from natural cotton as effective sorbents for oil spill cleanup. Ind Eng Chem Res 52:18251–18261

70. Bi H, Yin Z, Cao X, Xie X, Tan C, Huang X et al (2013) Carbon fiber aerogel made from raw cotton: a novel, efficient and recyclable sorbent for oils and organic solvents. Adv Mater 25:5916–5921
71. Li Y-Q, Samad YA, Polychronopoulou K, Alhassan SM, Liao K (2014) Carbon aerogel from winter melon for highly efficient and recyclable oils and organic solvents absorption. ACS Sustain Chem Eng 2:1492–1497
72. Cui Y, Xu G, Liu Y (2014) Oil sorption mechanism and capability of cattail fiber assembly. J Ind Text 43:330–337
73. Khan E, Virojnagud W, Ratpukdi T (2004) Use of biomass sorbents for oil removal from gas station runoff. Chemosphere 57:681–689
74. Cao S, Dong T, Xu G, Wang F (2016) Study on structure and wetting characteristic of cattail fibers as natural materials for oil sorption. Environ Technol 37:3193–3199
75. Sun X (2004) Acetylation of sugarcane bagasse using NBS as a catalyst under mild reaction conditions for the production of oil sorption-active materials. Bioresour Technol 95:343–350
76. Vlaev L, Petkov P, Dimitrov A, Genieva S (2011) Cleanup of water polluted with crude oil or diesel fuel using rice husks ash. J Taiwan Inst Chem Eng 42:957–964
77. Angelova D, Uzunov I, Uzunova S, Gigova A, Minchev L (2011) Kinetics of oil and oil products adsorption by carbonized rice husks. Chem Eng J 172:306–311
78. Bazargan A, Tan J, Hui CW, McKay G (2014) Utilization of rice husks for the production of oil sorbent materials. Cellulose 21:1679–1688
79. Uzunov I, Uzunova S, Angelova D, Gigova A (2012) Effects of the pyrolysis process on the oil sorption capacity of rice husk. J Anal Appl Pyrolysis 98:166–176
80. Chen P, Cho SY, Jin H-J (2010) Modification and applications of bacterial celluloses in polymer science. Macromol Res 18:309–320
81. Saibuatong O, Phisalaphong M (2010) Novo aloe vera–bacterial cellulose composite film from biosynthesis. Carbohydr Polym 79:455–460
82. Castro C, Zuluaga R, Putaux J-L, Caro G, Mondragon I, Gañán P (2011) Structural characterization of bacterial cellulose produced by *Gluconacetobacter swingsii* sp. from Colombian agroindustrial wastes. Carbohydr Polym 84:96–102
83. Sai H, Fu R, Xing L, Xiang J, Li Z, Li F et al (2015) Surface modification of bacterial cellulose aerogels' web-like skeleton for oil/water separation. ACS Appl Mater Interfaces 7:7373–7381
84. Ghosh A, Collie S (2014) Keratinous materials as novel absorbent systems for toxic pollutants. Def Sci J 64:209–221
85. Radetić MM, Jocić DM, Jovančić PM, Petrović ZL, Thomas HF (2003) Recycled wool-based nonwoven material as an oil sorbent. Environ Sci Technol 37:1008–1012
86. Radetic M, Ilic V, Radojevic D, Miladinovic R, Jocic D, Jovancic P (2008) Efficiency of recycled wool-based nonwoven material for the removal of oils from water. Chemosphere 70:525–530
87. Ifelebuegu AO, Anh Nguyen TV, Ukotije-Ikwut P, Momoh Z (2015) Liquid-phase sorption characteristics of human hair as a natural oil spill sorbent. J Environ Chem Eng 3:938–943
88. Jadhav AS, Naniwadekar MY, Shinde NH, Anekar SV (2011) Study of adsorbtion of oil from oily water using human hair. Int J Adv Eng Technol 2:37–51
89. Bhushan B, Chen N (2006) AFM studies of environmental effects on nanomechanical properties and cellular structure of human hair. Ultramicroscopy 106:755–764
90. Barone JR, Schmidt WF (2006) Effect of formic acid exposure on keratin fiber derived from poultry feather biomass. Bioresour Technol 97:233–242
91. Aguayo-Villarreal IA, Bonilla-Petriciolet A, Hernández-Montoya V, Montes-Morán MA, Reynel-Avila HE (2011) Batch and column studies of $Zn^{2+}$ removal from aqueous solution using chicken feathers as sorbents. Chem Eng J 167:67–76
92. Osamor A, Chinonyere P (2016) Oil spill clean up from sea water using waste chicken feathers. In: Fourth international conference on advances in applied science and environmental technology—ASET 2016. Institute of Research Engineers and Doctors, pp 61–64
93. Wang J, Geng G, Liu X, Han F, Xu J (2016) Magnetically superhydrophobic kapok fiber for selective sorption and continuous separation of oil from water. Chem Eng Res Des 115:122–130

94. Michel J, Fingas M (2016) Oil spills: causes, consequences, prevention, and countermeasures, pp 159–201
95. He J, Ding L, Deng J, Yang W (2012) Oil-absorbent beads containing β-cyclodextrin moieties: preparation via suspension polymerization and high oil absorbency. Polym Adv Technol 23:810–816
96. Hubbe MA, Rojas OJ, Fingas M, Gupta BS (2013) Cellulosic substrates for removal of pollutants from aqueous systems: a review. 3. Spilled oil and emulsified organic liquids. BioResources 8
97. Payne KC, Jackson CD, Aizpurua CE, Rojas OJ, Hubbe MA (2012) Oil spills abatement: factors affecting oil uptake by cellulosic fibers. Environ Sci Technol 46:7725–7730
98. Bayat A, Aghamiri SF, Moheb A, Vakili-Nezhaad GR (2005) Oil spill cleanup from sea water by sorbent materials. Chem Eng Technol 28:1525–1528
99. Lam SS, Liew RK, Cheng CK, Rasit N, Ooi CK, Ma NL et al (2018) Pyrolysis production of fruit peel biochar for potential use in treatment of palm oil mill effluent. J Environ Manage 213:400–408
100. Husseien M, Amer A abdel razik, El-Maghraby A, Taha N (2008) Experimental investigation of thermal modification influence on sorption qualities of barley straw. J Appl Sci Res 4:652–627
101. Hrnčič MK, Kravanja G, Knez Ž (2016) Hydrothermal treatment of biomass for energy and chemicals. Energy 116:1312–1322
102. Hammerschmidt A, Boukis N, Hauer E, Galla U, Dinjus E, Hitzmann B et al (2011) Catalytic conversion of waste biomass by hydrothermal treatment. Fuel 90:555–562
103. Karakasi OK, Moutsatsou A (2010) Surface modification of high calcium fly ash for its application in oil spill clean up. Fuel 89:3966–3970
104. Wang J, Han F, Liang B, Geng G (2017) Hydrothermal fabrication of robustly superhydrophobic cotton fibers for efficient separation of oil/water mixtures and oil-in-water emulsions. J Ind Eng Chem 54:174–183
105. Gert EV, Shishonok MV, Torgashov VI, Kaputskii FN (1990) Crystallization of amorphous cellulose in the form of inclusion compounds. J Polym Sci Polym Lett Ed 28:163–166
106. Sidik SM, Jalil AA, Triwahyono S, Adam SH, Satar MAH, Hameed BH (2012) Modified oil palm leaves adsorbent with enhanced hydrophobicity for crude oil removal. Chem Eng J 203:9–18
107. Wong C, McGowan T, Bajwa SG, Bajwa DS (2016) Impact of fiber treatment on the oil absorption characteristics of plant fibers. BioResources 11
108. Gulati I, Park J, Maken S, Lee M-G (2014) Production of carboxymethylcellulose fibers from waste lignocellulosic sawdust using NaOH/$NaClO_2$ pretreatment. Fibers Polym 15:680–686
109. Onwuka JC, Agbaji EB, Ajibola VO, Okibe FG (2018) Treatment of crude oil-contaminated water with chemically modified natural fiber. Appl Water Sci 8:86
110. Teli MD, Valia SP (2013) Acetylation of Jute fiber to improve oil absorbency. Fibers Polym 14:915–919
111. Sun X-F, Sun, Sun J-X (2002) Acetylation of rice straw with or without catalysts and its characterization as a natural sorbent in oil spill cleanup. J Agric Food Chem 50:6428–6433

# Carbon Based Materials for Oil Spill Treatment

**Neelima Varshney, Hammad Hasan, and Md Palashuddin Sk**

**Abstract** Accidental oil spills have become a focal point of research due to their profound impact on marine ecosystems and the environment. Oil contamination poses a significant threat to biodiversity and the ecological balance of our planet. Traditional oil removal methods are plagued by inefficiencies and drawbacks, necessitating the exploration of innovative approaches to address this critical issue. Materials endowed with superhydrophobic and superoleophobic properties have emerged as promising candidates in the pursuit of efficient separation of oil from oil–water mixtures. These materials exhibit exceptional absorption efficiency, recyclability, selectivity, and chemical inertness. Among them, carbon-based materials have proven to be the most promising contenders for oil separation from contaminated water. Their exceptional characteristics, including a large surface area, high surface area-to-volume ratio, porous nature, low density, remarkable mechanical and chemical stability, and eco-friendliness, make them the ideal choice for this purpose. This chapter delves into recent research on various carbon-based materials, including activated carbon, graphene, graphene oxide (GO), graphene aerogel (GA), graphite, carbon nanoparticles (CNP), carbon nanotubes (CNT), carbon nanofibers (CNF), and carbon membranes. It highlights their significant potential in efficiently cleaning oil spills from oil/water mixtures, offering environmentally sound solutions to this pressing issue.

**Keywords** Carbon-based material · Superhydrophobicity · Superoleophilicity · Carbon nanotube (CNT) · Carbon nanofiber (CNF) · Carbon nanoparticle (CNP) · Carbon dots (Cdots) · Graphite · Graphene · Graphene oxide (GO) · Graphene sponge · Graphene aerogel (GA) · Activated carbon · Carbon membrane · Oil spill · Contact angle · Ecosystem · Eco-friendliness · Selectivity · Recyclability · Absorbent · Reusability

---

N. Varshney · H. Hasan · M. P. Sk (✉)
Department of Chemistry, Aligarh Muslim University, Aligarh, Uttar Pradesh, India
e-mail: palashuddin.ch@amu.ac.in

© The Author(s), under exclusive license to Springer Nature Singapore Pte Ltd. 2025 
S. A. Bhawani et al. (eds.), *Biobased Materials and Their Composites for Oil Spill Treatment*, Composites Science and Technology,
https://doi.org/10.1007/978-981-96-6049-0_7

# 1 Introduction

Oil spill contamination is a recurring environmental crisis, stemming from oil extraction, storage, and transportation processes. Particularly, the transportation of oil via sea routes has witnessed numerous catastrophic accidents, causing extensive harm to marine life and ecosystems. Notable among these incidents are the Mexican Ixtoc oil spill (1979), the Exxon Valdez oil spill (1989), where an 11-million-gallon oil tanker ran aground on a reef, resulting in severe damage to coastal and marine environments, and the 'Deepwater Horizon oil spill' accident (2010), which resulted in the contamination of the Gulf of Mexico and the US coast. The Sundarbans oil spill accident (2014) scarred a UNESCO World Heritage Site, affecting the delicate food cycle of its mangrove ecosystem, aquatic organisms, and wildlife. Similarly, the East China Sea oil spill (2018) saw millions of tons of highly flammable fuel oil polluting fishing waters, while the MV Solomon Trader oil leakage (2019) wreaked havoc on the coral reefs in the Solomon Islands [1–5].

These alarming incidents underscore the grave consequences of seawater oil contamination. If unabated, it threatens to push numerous marine species to the brink of endangerment [5–7]. Beyond these high-profile oil spill accidents, oil pollution emanates from diverse sources, including oil refineries, lubricants, compressor condensates, petrochemical, textile, leather industries, and metal processing [8, 9].

Oil-containing wastewater, a byproduct of these industries, poses a substantial threat, containing toxic chemicals such as phenols and aliphatic and aromatic hydrocarbons, known to be carcinogenic and mutagenic to humans [6]. Historically, three primary methods—biological, mechanical, and chemical techniques—have been deployed to address these challenges [10–12]. The biological approach leverages microorganisms to degrade and convert large oil hydrocarbon chains into smaller, benign molecules, such as carbonic acid, water, and fatty acids [11–13]. Conversely, the mechanical method deploys various barriers, skimmers, and sorbents, both natural and synthetic, to combat oil spills [14–17]. In contrast, chemical methods employ dispersants and solidifiers to break down oil layers over water into minute droplets [18].

Among these techniques, absorbent materials have emerged as a highly effective means of oil removal and recovery. They offer advantages like low production costs, ease of synthesis, high selectivity, eco-friendliness, and extensive reusability [2]. Notably, surfaces modified to exhibit superhydrophobicity (water contact angle $>150°$) and superoleophilicity (oil contact angle $<5°$) have shown great promise in selectively absorbing hydrocarbons, various organic solvents, and oil while repelling water. These contact angles serve as key indicators of a material's oil- or water-repelling properties [19, 20].

Natural materials such as wool fiber [21], sawdust [22], and zeolite [23] have been harnessed for oil recovery from oily wastewater. While they possess attributes like a large surface area and low cost, their absorption capacity remains limited, coupled with issues of poor selectivity and recyclability. Therefore, an urgent need exists for the development of materials capable of overcoming these limitations.

Various techniques have been explored to produce superhydrophobic materials, encompassing polymer gels, natural fibers, organic hybrids, nano-composites, resins, and carbon-based materials. In this chapter, we will delve into the realm of carbon-based materials, showcasing their superiority among superhydrophobic materials due to their exceptional oil separation capabilities [1, 2].

# 2 Carbon-Based Materials: Advanced Solution for Oil Spill Recovery

In recent years, there has been a concentration of research on the development of highly porous polymer materials for the purpose of oil removal and recovery by using various types of techniques (Table 1). However, the synthesis methods for these polymers often prove to be neither cost-effective nor time-efficient. Additionally, the synthesis process emits harmful gases, contributing to environmental degradation. Conversely, carbon-based materials offer a compelling alternative thanks to their singular chemical and physical characteristics. These materials can maintain their integrity even under extreme temperatures and harsh environmental conditions, displaying enduring physiochemical stability [24].

The defining features of carbon-based materials, including superhydrophobic and superoleophilic surfaces, coupled with their high porosity and expansive surface area, position them as prime candidates for applications in oil spill removal [25]. These materials exhibit remarkable affinity for the oil phase, boasting high absorption capacity, exceptional selectivity, superior oil separation efficiency, and prolonged stability. Notably, carbon-based materials are environmentally friendly, straightforward to synthesize, and come at a relatively low production cost. Therefore, they are widely regarded as the ideal sorption materials [26, 27].

Presently, researchers are actively exploring various carbon-based materials for their potential as absorbents in oil spill removal, including carbon nanotubes (CNT), carbon nanofibers, carbon nanoparticles, graphene oxide, graphene sponges, and activated carbon (Fig. 1) [2, 28, 29].

## *2.1 Graphene*

The superhydrophobic nature of carbon-based nanomaterials proves to be very beneficial in the separation of oil and water. Some scientists have started fabricating superhydrophobic materials using graphene and its derivatives, which can effectively separate oil/water mixtures due to their low thickness and unique chemical and physical properties. The first report on graphene and its derivatives, as mentioned by Novoselov and Geim in 2004, brought attention to their unique features, such as high performance, functionality, a 2D structure, and graphene's hydrophobic properties

**Table 1** Types of techniques used in oil spill removal

| Techniques used for oil spill removal | Methods | Mechanism | Environmental effect | References |
|---|---|---|---|---|
| Physical | Skimmers | Mechanical removal | No adverse effect | Kleinguetl et al. (2020) |
| | Booms | Mechanical removal | Eco-friendly, no effect | Song et al. (2014) |
| | Manual removal | Utilize buckets, bailers, drums, trowels, and rakes | Not affected | Robertson et al. (1997) |
| | Adsorbent | Absorption | Natural sorbents are eco-friendlier | Singh et al. (2020) |
| Chemical | Solidifiers | | Not affected | Motta et al. (2018) |
| | Dispersants | Chemicals de-emulsified oil | Adverse effect on marine environment | Harvey et al. (1990) |
| | Emulsifiers | Demulsification | No harmful impact | Roulia et al. (2003) |
| Thermal | In situ burning | Burning | Causes pollution | Fritt-Rasmussen et al. (2015) |
| Biological | Microorganisms' bioremediation | Microbial degradation | Eco-friendly | Lin et al. (2010) |

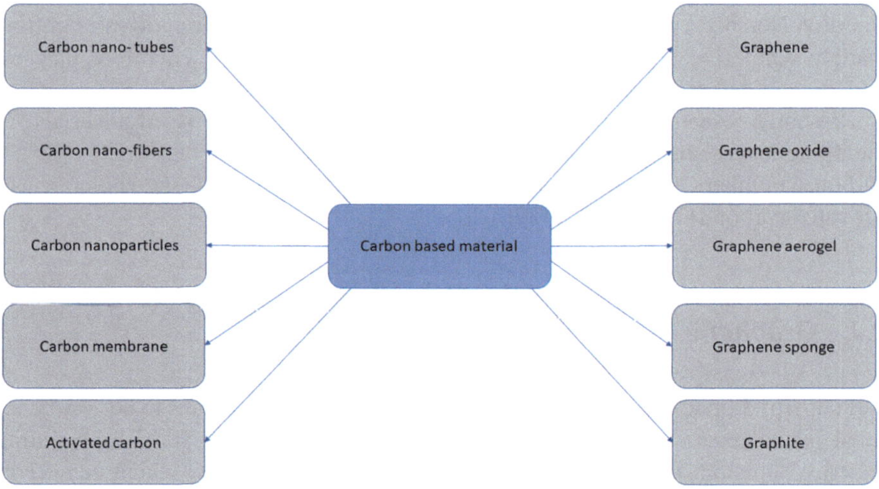


**Fig. 1** Different carbon-based materials implemented for oil/water separation

[30–34]. The atomic-level thinness of graphene channels greatly enhances separation processes. However, it has been observed that raw graphene, in its unmodified form, may not be as promising for various separation applications as other materials. Graphene-derived materials have garnered significant research attention for their potential in removing and recovering oil from oil-contaminated wastewater, which is generated by oil industries and during oil transportation. This is due to their exceptional mechanical, chemical, and thermal stability [35–37]. Graphene can be synthesized using various methods, including the reduction of carbon dioxide (CO2), reduction of graphene oxide (GO), chemical vapor deposition, mechanical exfoliation, and slicing of carbon nanotubes (CNT) [35].

Carbon forms a honeycomb structure to produce graphene, a two-dimensional sheet that appears to be just one atom thick. In 2004, a mechanical breakage method was employed to synthesize the first free-standing graphene. Currently, both top-down and bottom-up techniques are used for graphene synthesis. Top-down strategies yield large quantities of small-sized synthesized graphene sheets, whereas bottom-up procedures produce smaller quantities of larger-sized graphene sheets [35–37].

Graphene is an excellent choice for applications in oil sorption due to its exceptionally high surface area and its natural repellent properties.

Furthermore, 3D graphene, derived from 2D graphene nanosheets, offers numerous practical applications such as catalysts, capacitive deionization, lithium-ion batteries, sensors, sorbents, supercapacitors, microbial fuel cells, and vibration damping. It also finds applications in energy storage, catalysis, and energy conversion. 2D graphene nanosheets exhibit remarkable absorption capacity for toxic gases, certain water pollutants, oil, and even some toxic metals. These substances pose environmental concerns if left in an open environment [38, 39]. Furthermore, they are unsuitable due to the production of secondary pollutants, which are also harmful to society. Extensive research has been conducted on various sorption-related issues, including sorption kinetics and the recovery of oil, particularly using highly porous graphene materials [35, 39, 40].

Ge et al. reported that graphene-coated cotton (GCC) had an absorption capacity in the range of 30–50 times its own mass [41]. However, the absorption capacity of GCC is significantly lower compared to graphene-coated sponge (GCS) [42]. This is attributed to factors such as lower porosity, higher density, and increased hydrophobicity.

Sun and his research team demonstrated the remarkable properties of a 3D superwetting graphene-coated stainless-steel mesh. They calculated the water contact angle as 152°, and even when tilting the graphene mesh film to maintain the spherical shape of a water drop, the water droplet maintained this shape. This outstanding feature makes graphene mesh ideal for separating oil and water, making it suitable for oil spill removal and industrial applications [32, 43]. Abdala's group explores the potential of thermally reduced graphene (TRG) for oil spill cleanup. TRG, derived from graphite oxide, exhibits an exceptional sorption capacity, outperforming other carbon-based sorbents by absorbing up to 131 g of oil per gram of TRG. Moreover, with a filtration-based recovery process, just 1 g of TRG can remove roughly 300 g of crude oil across six cycles. TRG's performance depends on factors like pore volume, bulk density, carbon-to-oxygen ratio, and oil viscosity. While TRG selectively absorbs crude oil and is oleophilic, its sorption capacity diminishes slightly with repeated use [44].

## 2.2 Graphene Oxide

Oil spills are a significant environmental concern, impacting underground water, oceans, and nearby shores, while also resulting in the loss of vital energy resources. Research efforts are ongoing to find effective methods for recovering and utilizing spilled oil. Generally, there are three primary approaches to oil spill clean-up: mechanical, biological, and chemical. Mechanical approaches include techniques like application of sorbents and skimming. Chemical clean-up involves in situ burning, dispersion, and the use of solidifiers. Facilitating the microbial degradation of oil falls under the biological approach or bioremediation [44]. Among carbon-based sorbents, microporous graphene stands out for its exceptional oil-absorbing potential, offering a high oil yield per gram of sorbent. The efficiency of sorption of highly porous graphite relies on factors such as oil viscosity, temperature, density, and bulk sorptive density [45].

Graphene oxide (GO) can be synthesized by oxidizing graphene nanosheets, making it an attractive carbon-based compound thanks to epoxide, hydroxyl, and carboxylic acid moieties on the basal plane and edges of the nanosheets. These functional groups provide compatibility with polymers. When GO is incorporated into polymer matrices, it enhances mechanical strength and selectively restricts the passage of bigger molecules whilst permitting smaller ones to pass through. The hydroxyl and carboxylic groups on the GO's surface make it hydrophilic [46, 47].

There is an ongoing need for research in the field of GO-incorporated membranes that exhibit chemical, physical, mechanical, and thermal stability while remaining cost-effective. George Singhal, for instance, developed a novel super-hydrophilic and oleophobic graphene based membrane using a low-cost vacuum filtration technique. To enhance the surface and mechanical characteristics of graphene oxide (GO) sheets before creating the membranes, they were crosslinked using tannic acid (TA) molecules. This study demonstrated the high efficacy of TA crosslinked GO membranes in cleaning oily wastewater from industrial sources [48].

Abdalla and Wahab, using a phase inversion technique, manufactured polysulfone (PS) mixed matrix membranes (MMMs) with improved oil rejection, flux, and fouling resistance. These membranes incorporated functionalized graphene oxide (fGO) made from aspartic acid (AA). Their research showed that even with minimal fGO loading, the membrane's ability to exclude oil while allowing water transmission increased. PS/fGO MMMs can be effectively used for oil–water separation [49].

Liu and Wu developed a remarkably efficient graphene oxide (GO) membrane for extracting oil-in-water emulsions using a simple coating process. They connected GO nanosheets to a mixed cellulose ester membrane (MCEM) with a polydopamine

(PDA) layer through self-polymerization. This resulted in the creation of GO/PDA/MCEM, which successfully separated oil-in-water emulsions [50].

Gao et al. synthesized graphene oxide foam through various freezing-drying methods and improved its absorption capacity for solvents by attaching 1° amine-functionalized GO onto polyurethane sponges using nitrile groups [51]. Niu's group developed a reduced form of porous graphene oxide through a leavening process using an autoclave, which exhibited exceptional material weight capacity [52].

This research offers a simple, cost-effective method for creating an effective graphene oxide based material for oil-in-water emulsion separation. In conclusion, graphene oxide and its composites hold great promise for efficient oil cleanup thanks to their high sorption potential, promising recyclability, and selectivity for oil.

## *2.3 Carbon Nanotubes (CNT)*

Carbon nanotubes, a significant member of the carbonaceous materials, possess unique properties that have driven extensive research. The golden age of carbonaceous materials research began nearly two decades ago with the 1991 CNT report [53]. Carbon nanotubes hold great promise for oil–water separation due to their remarkable mechanical strength, quick rates of sorption, high capacity, and customizable surface chemistry [54–57]. Figure 2 illustrates the structures of single-walled CNT (SWCNT) and multi-walled CNT (MWCNT), showcasing their intricate designs.

Various methods, including laser ablation [58], arc discharge [59], pyrolysis [60], electrochemical approaches [61], and chemical vapor deposition [62], have been employed to synthesize carbon nanotubes. For example, Sun and team developed p-phenylenediamine modified carbon nanotubes (P-CNTs) through a diazotization reaction. These P-CNTs exhibited increased surface roughness, reduced surface free energy due to organic molecule introduction, a high mesopore ratio, a pore volume of 0.65 $cm^3/g$, a mesopore volume of 0.64 $cm^3/g$, and a surface area of 285 $m^2/g$. They displayed excellent hydrophobicity with a water contact angle of 140.8° and could absorb 3 to 12 times their mass in oil and organic solvents while repelling water. Substances absorbed could be easily recovered through heat treatment or solvent washing, though absorption capacity slightly decreased with reuse due to residual substances [63]. Ong and his team introduced an innovative method to augment the oil/water selectivity of porous poly(dimethylsiloxane) (PDMS) by attaching carbon nanotubes (CNTs) using sugar as a temporary template. This cost-effective modification significantly improved absorption capacity and speed compared to pure PDMS absorbents. The enhancement was attributed to CNT incorporation, which roughened the surface at the nanoscale, reduced surface energy, and increased hydrophobicity and oleophilicity, enabling selective oil absorption while repelling water [64]. Piperopoulos et al. developed a composite foam containing carbon nanotubes (CNTs) within a silicone PDMS matrix and a Sn-based catalyst. They tested these foams with both pristine and functionalized CNTs in various oils. The foam with pristine

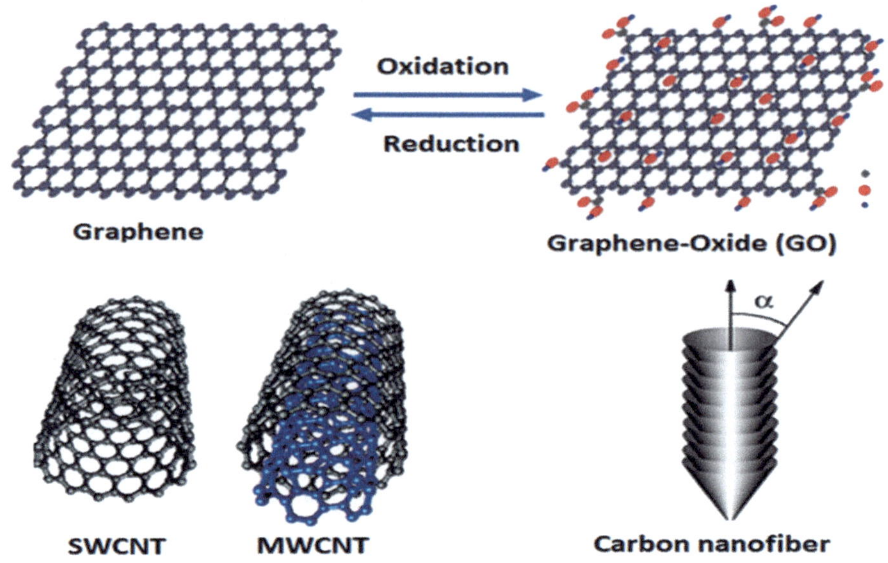


**Fig. 2** Structure of Graphene, Graphene Oxide, SWCNT, MWCNT and Carbon nanofiber. Reproduced from the reference [74]

CNTs exhibited the greatest sorption capacity (800wt% in virgin naphtha) and could be reused up to 10 times [65]. Gui et al. produced magnetic CNT sponges and carbon nanotube sponges using chemical vapor deposition (CVD). These sponges had interconnected porous structures with varying pore sizes, outstanding mechanical properties, resistance to cyclic stress in oils, and chemical stability in various environments. Both types of sponges were hydrophobic, showing water contact angles measuring 156° and 140°. The CNT sponge had lower density and better hydrophobicity, resulting in higher oil and solvent absorption capacity (80–180 times its weight). Remarkably, it could be reused over 1000 times without significant changes [66, 67]. Lee et al. employed thermal chemical vapor deposition with an $Al_2O_3$ film as a diffusion barrier to synthesize vertically-aligned carbon nanotubes (CNTs) on a stainless-steel substrate. This resulting CNT-stainless steel mesh functions as an effective filter for both organic solvents and oil, exhibiting a 150° water contact angle and blocking 80% of water. Heating the mesh enhances the rate of flow of viscous emulsions; however, it remains porous enough for fine water droplets to pass through [68]. Gea et al. presented a magnetically superhydrophobic bulk material with an impressive absorption capacity ranging from 0.71 to 1.15 times its own weight. This substance was fabricated through a process involving the hot pressing of PTFE, multiwall CNTs, and iron oxide. It showcased superhydrophobic properties on both its surface and interior, boasting a 158° water contact angle. Notably, the absorbed oil could be efficiently removed through combustion, and the material maintained thermal stability even after undergoing ten cycles of oil removal [69]. Liu et al. introduced a highly effective method for separating emulsified oil from wastewater using functionalized multiwalled carbon nanotubes (F-MWCNTs) as demulsifiers. They achieved an outstanding 99.8% oil removal within minutes, reducing the initial oil concentration in the water from 1741.2 mg/L to a mere 56 mg/L. The separated oil phase exhibited a low water content of 0.45%, satisfying transportation and refinery standards. This remarkable success was attributed to F-MWCNTs breaking down the protective film on oil droplets, facilitated by strong interactions with asphaltenes/resins via p–p and/or n–p interactions [70]. Wang et al. created CNT-enhanced polyurethane sponges with superhydrophobic and superoleophilic characteristics. They absorb oils rapidly, are mechanically strong, and can be reused without losing their oil-absorption power up to 150 cycles [71]. Gu and team prepared a silver/polyacrylic acid (PAA) CNTs (Ag/PAA CNTs) composite membrane for separate three phase system of solid/water/oil. CNT and its derivatives prove as the best candidates of reusable sorbent for oil from water and shows oil absorption and reusability with a large extent [72].

Researchers have further explored modifications that enhance pore size and absorption capacity. Presently, the development of a polymer surface covalently bonded with CNTs shows promise for efficiently separating oil/water emulsions, even when stabilized by surfactants [72, 73]. CNTs and their derivatives continue to emerge as leading candidates for reusable sorbents in oil–water separation, demonstrating substantial oil absorption capacity and reusability.

## 2.4 Carbon Nanofiber

In the realm of oil–water separation, nanofibrous materials are currently gaining prominence as effective sorbents [75]. Various types of fibers, including cellulose fiber functionalized with acetate [76], cotton thread [77], and polymer fibers [78], have been synthesized and demonstrate the ability to absorb oil. The electrospinning method (a top-down approach) is commonly used to fabricate highly porous, chemically inert, and uniformly-sized organic or biological fibers [79]. However, one challenge with these fibers is their tendency to dissolve in solvents or swell during the recycling process of oil absorption.

Carbon-based nanofibers (CNF) address this issue effectively. Figure 3 illustrates the Modification of CNT into CNF and CNF to Carbon fiber. Liu et al. prepared CNF using terephthalic acid (PTA) with sublimating agents. These CNFs possess macropores, which enhance flexibility, reduce stress, and exhibit significant hydrophobicity and oleophilicity. Macro-porous carbon nanofibers (MCNF) have a high oil absorption capacity (138.4 g/g with silicone oil) and an 89.2% degree of porosity. MCNF can extract oil by forming a membrane using organic solvents and can be successfully reused up to 5 times without significant oil loss, although there is a potential for secondary pollution [80]. Tai et al. discovered carbon nanofiber composites with silica, which exhibit greater flexibility and toughness compared to pristine CNF. Silica-composite CNF, after silicone coating, demonstrates ultra-oleophilicity and superb hydrophobicity, characterized by a water contact angle of 144.2°. These

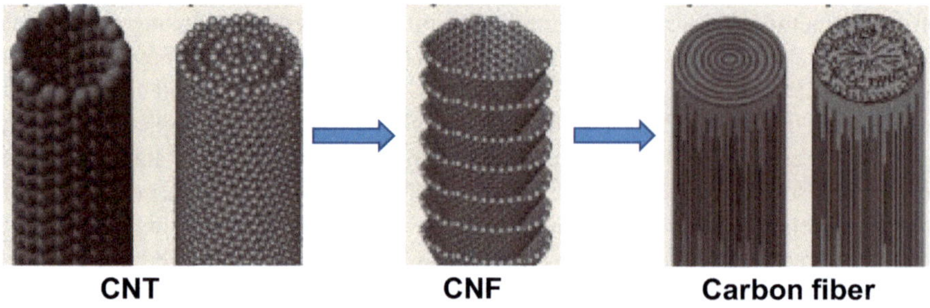


**Fig. 3** Modification of CNT into CNF and CNF to Carbon fiber. Reproduced from the reference [91]

composites remain stable at extreme pH levels and high temperatures. However, one drawback of CNF is the time-consuming nature of oil recovery [81]. Wu and colleagues synthesized ultralight, highly flexible, and cost-effective CNF aerogels through the pyrolysis of bacterial cellulose. These aerogels have an exceptionally low density of about 4–6 mg/cm3 and can withstand manual pressure. They prove to be efficient sorbent materials for oil, with a remarkable absorption capability of as much as 310 times their own mass [82].

CNF, when combined with various porous substrates, can yield a material remarkably effective for oil–water separation. Polyurethane, inherently porous, undergoes a transformation when coated with a CNF network, resulting in a hydrophobic material with exceptional absorbent properties. This combination proves effective in separating oil and water, thanks to the spongy nature of the foam, which can return to its original size after oil retrieval. Notably, it possesses small pore sizes, a large surface area, low surface energy, and capillary action, all essential for efficient oil absorption. This method is cost-effective in terms of raw materials and yields a material with enhanced mechanical and chemical strength. It was originally developed by Baig et al. more than three decades ago [83]. Another approach involves the creation of carbon nanofiber (CNF) sponges through a straightforward chemical vapor deposition method, using $C_2H_4$ as a source of carbon and a Ni-Cu alloy as a catalyst. These sponges feature a three-dimensional network structure, with thick CNFs as reinforcement rods and thin CNFs as binding wires. This unique structure provides impressive mechanical flexibility and high compressive strain tolerance ($> 90\%$). Moreover, they showcase super hydrophobicity (with a contact angle >153°) and high oleophilicity. These CNF sponges can absorb oils at remarkable rates, ranging from 22 to 75 times their weight, and their absorption capacity can be controlled by adjusting density or porosity. Additionally, these sponges are highly recyclable through mechanical squeezing. Overall, CNF sponges exhibit great potential for various environmental protection applications due to their unique properties as reusable sorbent materials for oil absorption [84]. Xiao and colleagues employed coal liquefaction residue to fabricate a highly effective composite material comprising carbon nanofibers integrated with a carbon foam base. This composite displayed remarkable absorption capabilities, with the ability to absorb oils and organic solvents at rates between 15 and 28 times its own weight. Their method involved creating carbon foam using template synthesis and growing carbon nanofibers on the foam through catalytic chemical vapor deposition, with iron moieties in the coal liquefaction residue acting as catalysts. This resulted in a network of carbon nanofibers covering the entire foam surface, increasing surface roughness and enhancing hydrophobicity, evident in a water contact angle of approximately 140° [85]. A recent study introduces a simple and eco-friendly method for producing large-scale ultralight carbon nanofiber (CNF) aerogels from bacterial cellulose (BC) pellicles. These CNF aerogels arc fire-resistant, highly compressible, and exhibit remarkable efficiency in absorbing organic solvents and oils, with the capacity to absorb up to 310 times their weight. Additionally, the CNF aerogels show sensitivity to compressive strain, making them suitable for pressure-sensing applications [86]. Yu et al. developed CNF-derived aerogels and hydrogels through hydrothermal carbonization using glucose and tellurium. The resulting fiber diameter can be adjusted by changing the concentration of glucose and reaction time [25, 87]. This method initially produces a hydrophilic fiber due to the presence of a hydroxyl group on the surface. To achieve a hydrophobic surface, treatment with poly-dimethyl siloxane (PDMS) [88] is required. This treatment enhances strength, flexibility, corrosion resistance, oil absorption capacity, and photothermal conservation capacity while increasing the number of capillary tubes to enhance capillary action. The material showcases superhydrophobicity, with a greater than 150° water contact angle. These CNF composites effectively absorb various types of oil and maintain their properties even under high temperatures and extreme pH conditions. Furthermore, in the presence of light, the viscosity of crude oil decreases, enabling efficient oil absorption from the sea surface in harsh conditions [89]. G-CNF foam has proven to be a suitable composite for oil spill removal. A 2-dimensional electrospun poly-acrylonitrile itaconic acid (PANIA) sheet was transformed into a self-assembled network of 3D-Aerogels using a gas foaming method. This provides a simplistic and versatile approach to creating 3D CNF composites with multi-layer graphene. Polyurethane can be coated with graphene oxide (GO) followed by carbonization to produce 3D G-CNF. These GO and CNF composites exhibit synergistic properties, enhancing surface area-to-volume ratio, porosity, good chemical stability, flexibility, thermal and mechanical stability, and fire resistance. They can absorb oil at capacities ranging from 86 to 153 g/g, and the material remains recyclable for up to 11 cycles during the oil absorption process [90]. In recent years, numerous reports have focused on CNF and its modified forms for effective and expert oil spill removal.

## *2.5 Carbon Membranes*

Numerous membranes made from different polymers have been explored for separating oil and water mixtures. However, they often face limitations such as low chemical, thermal, and mechanical stability, along with a lack of recyclability [92,

93]. In response to these challenges, current research has turned toward economically affordable and environmentally friendly carbon fibers for immediate oil removal from water surfaces. These carbon fibers are assembled into membranes, capturing the attention of researchers for further investigation. This carbon-based membrane offers several advantages, including cost-effectiveness, higher chemical inertness, mechanical strength, ease of recovery, and the ability to recycle oil [6]. Carbon nanomaterial-containing membranes have proven to be relevant for oil recovery and removal treatments, demonstrating sustainability in aggressive and harsh environments, such as high temperatures, extreme pH levels, high pressure, vaporized solvents, and the presence of oxidizing agents. Over the years, numerous carbon-based membranes have been reported for treating oil-containing wastewater. Li et al. developed a carbon-based membrane with promising potential for filtering oily wastewater and effectively recovering oil. Their findings indicated a pore size of 10 mm, an optimized pressure of approximately 0.1 Megapascal, and a cross-flow velocity of 0.1 m/sec for efficient sieving. The filtration rate depends on pore size, applied pressure, and cross-flow velocity. This carbon-based membrane deforms when the pore size is too large or the applied pressure is excessive, resulting in decreased filtration flux. However, a high cross-flow velocity promotes recovery. To enhance antifouling behavior and absorption efficiency, coal was incorporated as an anode into the membrane, interacting with the electric field of the carbon membrane, which improved electrical conductivity and made it an auspicious choice for filtering oily wastewater and recovering oil. The positive impact of coal on the membrane also reduced degradation and depended on the amount of coal, as an excessive amount of oil droplets increased the rate of membrane degradation and oil drop collapse. This carbon-based membrane boasts excellent regeneration capacity, enabling both chemical and physical cleaning. The cleaning process involved immersing the membrane in deionized (DI) water for 15 min at room temperature, followed by soaking it in a solution of 0.1 M sodium hydroxide and 0.2 g/L sodium dodecyl benzene sulphonate (SDBS) for 60 min at 60 °C. Finally, it was immersed in a 0.1 mol/L HCl solution for another 60 min, followed by repeated washing with DI water for 15 min. After this process, the efficiency of oil removal reached 95%, with a recovery flux of 88.4%. The values before and after cleaning the membrane were nearly the same [94]. Similarly, Wu et al. determined the concentration of oil, additives, operating time, and the efficiency of regeneration for oil removal from oil wastewater employing a self-prepared microfiltration membrane made from phenolic resin. The membrane's pores allowed for the filtration of oil while restricting the passage of water, resulting in an exceptionally high flux of $10^5$ $L/m^2$ h bar. This membrane was able to filter oily wastewater at least twice or thrice as effectively as commercial methods, achieving over 99% oil separation efficiency. Two factors contributed to this outstanding efficiency: the nanometer-thin thickness of the SWCNT film membrane with the same pore size as commercial membranes, which are typically twice or thrice as thick, and the super oleophilic property of the SWCNT film. When an oil–water emulsion contacts this membrane, it undergoes de-emulsification, allowing oil to rapidly collect on the membrane's surface due to its strong oil affinity and water-repelling properties [95]. Hsieh et al. developed a membrane composed of carbon fabric (CF) and CNT. The CNT-CF membrane achieved an exceptional 99.7% efficiency in separating oil and water. Thinner CF membranes and thinner CNT-CF membranes resulted in higher filtration rates. The oil permeability of the CNT-CF membrane reached 180 L/m$^2$h bar, significantly better than CF membranes without CNT, which measured only 95 L/m$^2$h bar. This improvement was attributed to the kinetic energy of oil droplets on the membrane surface and the superhydrophobic property characterized by a 165° contact angle with water droplets [96]. Shi et al. found that the water flux through the SWCNT film was much lower than that through the CNT-CF composite membrane. The original CF surface displayed both hydrophobic and oleophobic properties, which could be converted to hydrophobic through fluorine treatment, effectively reducing surface energy. This carbon-based membrane exhibits oil absorption capability and regeneration ability, resulting in minimal mass loss [9]. The use of carbon-based membranes holds promise for addressing various environmental challenges due to their carbon content, cost-effectiveness, and high efficiency with minimal energy loss.

## 2.6 Graphene Aerogels

Graphene aerogel (GA) stands out as a prominent aerogel for the capture of oil from oil–water mixtures. The initial material used in the synthesis of graphene aerogel is graphene oxide (GO). The production of GA involves two crucial steps which include freeze-drying, that eliminates organic functionalities from the surface and enhances the carbon-to-oxygen ratio. Consequently, it has been observed that GA exhibits an absorption capacity 40 times greater than that of GO, achieved by weight of the absorbent. The remarkable mechanical strength and fire resistance of GA render it highly suitable for the absorption of organic solvents, boasting a capacity of 97%. Importantly, GA maintains this performance even through up to 10 cycles of absorption and compression when dealing with bean oil [97]. Furthermore, GA can be synthesized through a process involving ethylene diamine (EDA) ammonia via double hydrothermal reduction, yielding a graphene hydrogel followed by a freeze-drying step. This method enhances the sorption capacity, resulting in 130 gm/gm for diesel oil and 71.67 gm/gm for oil emulsion. Improved hydrophobicity arises from the reduction of oxygen-containing functionalities, while enhanced mechanical properties increase its practicality across various fields. It's worth noting that GA synthesized via this method can withstand up to 50% strain before damaging the graphene sheet's structure, limiting oil recovery to 50% after each compression. Continual experimentation indicates that only 6.4% of oil can be retrieved after 15 cycles. Additional research has reported on the preparation of GA, including a method involving one-step GO reduction with vitamin C under optimized conditions, resulting in a material with excellent absorption capacity. Key parameters affecting the structure and oil absorption capacity include the size of GA, carbon particle characteristics, the GO-to-Vitamin C ratio, GO concentration, and solution pH. Achieving optimal conditions, including a 500-mesh C-size, a specific GO-to-Vitamin C ratio, a 2.5 mg/ml GO concentration, and a neutral pH, enables a capacity range of 138–328 gm/gm. Impressively, the absorption capacity remains consistent through 10 cycles of repeated absorption and compression, maintaining an efficiency of 92% [98]. Another study employs GO, formaldehyde, and resorcinol as reactants for GA synthesis using the conventional sol–gel method. Notably, GA prepared through this process exhibits reduced sensitivity to temperature. For instance, at –40 °C, it achieves a capacity of 26.5 gm/gm, and even at temperatures up to 40 °C, it maintains a capacity of approximately 25.0 gm/gm. However, if the temperature rises to 240 °C, the capacity reaches only 24.7 gm/gm with motor oil. It's important to highlight that increased temperature results in reduced capacity, accompanied by alterations in the viscosity of the motor oil. Intriguingly, temperatures of 80 °C or above lead to the disappearance of the oil layer [99].

Presently, extensive research has been conducted on the preparation, characterization, and application of GA. GA obtained through the bidirectional freeze-casting method exhibits a comparatively higher number of capillaries, facilitating the absorption of organic liquids. The sorption capacity falls within the range of 86.62–374.7 gm/gm, and importantly, there is no significant loss of reusability even through 10 cycles of absorption and pressing. Well-aligned GA, obtained through unidirectional freezing or bulk freezing of reduced GO using hydrothermal methods, exhibits improved hydrophobicity with increased hydrothermal time. The result is a reduction in water and oxygen content within the GA hydrogel, rendering the material more hydrophobic. Specifically, the absorption capacity of GA synthesized through this method for n-Hexane reaches 105 gm/gm, and this performance remains consistent through 20 cycles [100]. Moreover, a biomimetic GA structure, fabricated through a systematic combination of freeze-drying and unsaturated reduction using water vapors, demonstrates an absorption capacity of 55 gm/gm with 90% efficiency after 10 cycles of reusability [101]. Importantly, the pore size of GA influences the absorption rate, although it does not impact the equilibrium adsorption capacity. Consequently, from this comprehensive review, it becomes evident that graphene aerogels hold significant potential for playing a crucial role in the treatment of oily wastewater.

## 2.7 Graphene Sponge

Numerous research studies have highlighted the potential of three-dimensional, porous sponges as high-capacity sorbents due to their substantial surface area. Although, most commonly produced sponges suffer from poor hydrophobicity, limiting their effectiveness in oil spill recovery and remediation [102]. To address this limitation, surface modifications have been employed to enhance the hydrophobicity and reduce water-wettability of sponges. Various coatings, including carbon nanotubes, graphene, nanofibers made of polyaniline, polypyrrole, and nanocrystals, have been frequently utilized for this purpose [103]. The graphene-based sponge, owing to its extensive surface area, exhibits exceptional sorption capabilities for various types of oil [104].

In a study conducted by Rahmani and Samadi, graphene-coated polyurethane (PU) sponges were investigated for selective oil spill removal from water. To prepare nanoporous graphene polyurethane sponge (NPG/PU) and graphene oxide polyurethane foam (GO/PU), granules of graphene oxide (GO) and nanoporous graphene (NPG) were compared with highly porous PU. The researchers found that NPG/PU sponges effectively removed organic liquids, including crude oil stains, outperforming GO/PU sponges. Graphene-coated polyurethane (PU) sponges, therefore, offer an excellent and low-cost solution for oil spill cleanup [105]. Additionally, Xia and their team successfully developed a superhydrophobic reduced graphene oxide-coated polyurethane (RGO@PU) sponge using a one-pot solvothermal approach in the ethanol presence. This RGO@PU sponge featured a superhydrophobic surface and displayed a high oil-absorption capacity, surpassing its initial mass by 37 times. With an impressive oil/water separation efficiency of 99% and a rapid absorption rate, RGO@PU sponges demonstrate significant potential for use as oil/water segregation absorbents [106]. Pethsangave and co-workers, using a one-pot hydrothermal technique, introduced an innovative superhydrophobic ultra-light graphene-based carrageenan sponge (GCS) absorbent. This sponge is designed to selectively capture organic solvents and oils from water mixtures. Incorporating carrageenan into graphene oxide (GO) nanosheets, followed by reaction with formaldehyde, resulted in a hydrophobic and cross-linked structure. The GCS exhibited superior oil absorption capacity, ranging from 25.2 to 50 g of oil per gm of adsorbent, highlighting its potential for efficient oil removal from water [107]. In another report by Bi and his research group, a graphene sponge with rapid absorption rates of $0.57^{-1}$ gm/sec and an absorption capacity ranging from 20 to 86 times its de-oiled material was presented. This was achieved through the reduction of graphene oxide in suspension and subsequent molding into a spongy graphene material with numerous capillaries. Iqbal's group [44] synthesized thermally reduced graphene (TRG), whose absorption capacity was tunable through variations in exfoliation conditions, pore volume, bulk density, and the carbon-to-oxygen ratio. The maximum sorption capacity of 131 gm/gm was achieved at a C/O ratio of 17:1. Wang and his team [108] reported a green methodology-based reduction of graphene oxide using phenolic acid obtained from natural sources, with gallic acid showing excellent results due to its porous and mechanically robust behavior. This graphene exhibited an oil absorption capacity of 15–61 and a dye absorption capacity of 115–1260 mg/gm.

Overall, graphene-based materials composed of carbon have demonstrated desirable qualities, including mechanical stability, high capacity, high porosity, and substantial porous volume [107]. The synthesis of carbon materials with increased hydrophobicity, oleophilicity, and specialized thermal properties has garnered significant attention [109]. Graphene sponges fulfill these properties to a significant extent.

## 2.8 Graphite

Exfoliated graphite stands as a noteworthy carbon-based material for mitigating oil/water contamination. It exhibits the capability to rapidly absorb heavy oil from water surfaces, even at temperatures as high as 1000 °C. This remarkable conversion of natural graphite into a porous material is achieved through a microwave-assisted method [110] or a process involving chemical intercalation followed by thermal treatment.

In one study by Bayat's research group, a gram of material was synthesized through a thermal shock technique involving nitric acid and sulfuric acid intercalation with graphite [111]. Hristea's group pursued a similar approach, creating a sorbent by intercalating sulfuric acid with graphite and inducing thermal expansion. Graphite oxidation was facilitated using either potassium permanganate ($KMnO_4$) or ferric chloride ($FeCl_3$) as oxidizing agents. $KMnO_4$ exhibited superior results due to its high degree of oxidation and the resulting significant volume expansion. Nevertheless, oxidation with FeCl3 is considered more practical due to its less exothermic reaction and ease of filtration. The surface area-to-volume ratio of the pores and the oil absorption capacity of exfoliated graphite are dependent on the ratio of $H_2SO_4$ to $HNO_3$. Absorption capacity through exfoliated graphite came to be 2.4–9.7/100 gm [112]. Toyoda and colleagues conducted investigations into the absorption capacity of industrially synthesized and commercially obtained exfoliated graphite under varying pressure and temperature conditions. The application of different pressure intensities can alter pore volume, density, and absorption capacity. Notably, the absorption capacity tends to decrease with decreasing temperature and viscosity. Maximum oil absorption capacity reached 80–90 g/g of absorbent, with approximately 80% oil recovery achieved through pressing or suction methods [113]. Hu and colleagues successfully engineered a graphite composite using isoprene and isobutylene rubber cryogels, featuring an intricate porous network crosslinked with sulfur monochloride through the cryogelation technique. Notably, the surface exhibited fracturing due to the crosslinking of the graphite. The material's water-repellent properties and absorption capacity were found to be contingent on both temperature and the concentration of graphite [114].

The stability of the pores was observed to persist below 0 °C, although there was a notable drawback: the rapid nucleation rate led to a reduction in pore volume. To achieve maximum absorption, it was deduced that an optimized quantity of graphite was essential, as an excessive amount of graphite corresponded to diminished absorption capacity. When the concentration of cryogels was set at 2% w/w of graphite at −15 °C, the material exhibited rapid oil removal capabilities, removing between 17.6 and 23.4 g of various oils. Wang's research group introduced a novel nanomaterial for oil and lubricant absorption known as Modified Expanded-Graphite (MEG). The synthesis involved the reaction of graphite with strong acids such as $H_2SO_4$ and oxidants like $H_2O_2$. MEG displayed a distinctive warm-like structure, characterized through FESEM (Field Emission Scanning Electron Microscopy) analysis, revealing three distinct types of pores: V-shaped, willow leaf-like, and pores formed by the up-and-down movement of pore walls. Notably, the first two types of pores exhibited an affinity for storing various oils. As synthesized, MEG demonstrated an impressive oil absorption capacity ranging from 43.25 to 84.68 g/g of absorbent, with a corresponding volume expansion falling within the range of 100–320 ml/g [115]. In light of these findings, it is evident that graphite, as a carbon-based material, holds significant promise for oil absorption applications, making it suitable for various oil–water separation systems capable of effectively separating both liquids as shown in Fig. 4.

**Fig. 4** Oil removal with the help of powdered graphite

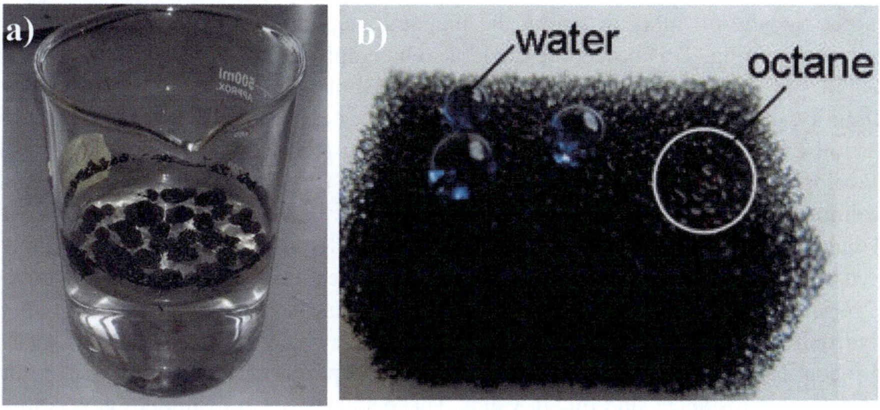


**Fig. 5** **a** Activated carbon used in oil removal. Reproduced from the reference [117] **b** Activated carbon-coated sponge show superhydrophobicity. Reproduced from the reference [119]

## 2.9 Activated Carbon

Surface water quality is often compromised by various contaminants, with oil contamination being a significant concern for both surface water and seawater. Figure 5a depicts the oil removal capability of activated carbon, while Fig. 5b showcases the superhydrophobic properties of activated carbon-coated sponge. Utilizing activated carbon derived from waste materials, such as maize cobs, offers a promising approach for mitigating oil spills in aquatic environments. Shokry et al. employed an eco-friendly approach by using magnetic activated carbon pellets synthesized from hyacinth plants for oil adsorption. This innovative approach transformed water hyacinth waste biomass into a cost-effective nano-magnetic adsorbent (NMAC) suitable for efficient oil spill cleanup. Activated carbon extracted from the plant's roots was combined with nano-magnetite to create NMAC. Remarkably, NMAC exhibited a high oil adsorption capacity of 30.2 g oil/g and could remove 80% of oil within 60 min when used at a concentration of 1 g/L [116]. In another study, coconut-coir activated carbon (CCAC) was treated with potassium hydroxide (KOH) to enhance its efficiency in removing crude oil from water. Increasing contact time, adsorbent dose, and initial oil concentration improved oil removal. The modified CCAC-KOH demonstrated a remarkable adsorption capacity of 4859.5 mg/g at 304 K, making it a promising, cost-effective, and eco-friendly choice for oil spill cleanup [117]. Joy and colleagues synthesized activated carbon from coconut shells and combined it with superparamagnetic iron oxide nanoparticles to create a magnetic nanocomposite. This research showcased the composite's effectiveness as a material for magnetic separation-based oil spill cleanup. The use of magnetic interactions for recovering the composite-oil mixture makes the magnetic separation method quick and less time-consuming. This procedure could also be employed to contain oil shortly after a spill, preventing its spread. With additional advantages such as affordability, recyclability, ease of use, and a scalable synthesis process, this material holds promise for oil remediation when combined with existing technologies [118]. Deng et al. developed highly porous activated carbon with an extensive surface area and pore volume by subjecting commercially available activated carbon to KOH activation. Treatment with polydimethylsiloxane (PDMS) rendered the activated carbon superhydrophobic, with a 163.6$^{o}$ water contact angle. It exhibited impressive absorbency (2695–8586 wt.%) for various organic compounds and oils. Additionally, superhydrophobic activated carbon-coated sponges were created using a straightforward dip-coating method, demonstrating excellent absorption selectivity for a wide range of organic substances and oils in water. These sponges also exhibited recyclability and hold significant potential for large-scale removal of organic contaminants and oil spills from water, opening up possibilities for various applications, including oil spill cleanup, liquid–liquid separation, and water treatment [119]. The unique properties of carbon adsorbents derived from biomass, such as mechanical integrity, a large internal surface area, and regenerability, have led to their widespread use in the removal of hazardous chemicals from the environment. Environmental concerns have driven the need for carbon adsorbents to sequester heavy metals, dyes, and components of crude oil. This has prompted studies on carbon adsorbents made from agricultural and domestic biomass waste. Activated carbon sourced from biomass is highlighted for its effectiveness in removing toxic impurities from water solutions, including oil spills, dyes, and various hazardous compounds [120].

### 2.9.1 Carbon Nanoparticle

Highly porous carbon materials with superhydrophobic properties have garnered significant attention in research owing to their prospective applications in water treatment, water/oil separation, and oil spill cleanup. However, most reported superhydrophobic carbon materials involve complex synthesis procedures requiring catalysts and high temperatures, limiting their practicality. Dai and his colleagues have introduced a superhydrophobic carbon-based porous material known as carbon nanoparticles (CNP). Superhydrophobicity is characterized by a water contact angle of 150°. This material is synthesized using $C_2H_2$ and a mixture of acetylene and carbon tetrafluoride gas, generated by luminescence discharge of plasma at room temperature, without the need for any catalysts. The low surface energy responsible for the superhydrophobic properties is attributed to the fluorine content. The surface structure depends on the ratio of carbon tetrafluoride to acetylene in the gaseous mixture and the applied pressure. These particles are capable of separating oil–water pairs so it is used for deoiling from oily water effectively [121]. CNP can be synthesized using both top-down and bottom-up techniques. Recently, the bottom-up approach has gained relevance, involving the use of chemical and natural precursors in suitable solvents. Shi and her team utilized ultrasonication technology to prepare CNP-modified foam. Their work showed that CNP strengthened the polyurethane (PU) foam and increased surface roughness, likely due to the formation of a strong bond between PU and CNP. The bond strength between PU and CNP was found to be comparatively stronger than that of carbon nanotubes (CNT) and graphene oxide (GO) foam. This carbon-based material exhibited excellent mechanical strength, even after undergoing 500 cycles of squeezing for oil recovery without significant deformation. The water contact angle for CNP-PU was reported as 127.6°, indicating strong water repellency. It has the capacity to absorb 50–120 times its absorbent weight [122]. From the review, it can be inferred that tetraethyl orthosilicate (TEOS) plays a crucial role in creating holes in the carbonized polysulfone beads. The collapse of the beads is prevented by the presence of silica powder, and the quantity of silica also influences the capacity for oil and organic solvent absorption. These carbon beads can absorb 50–55% of their own weight, and their recyclability is maintained for up to 6 cycles [122]. Carbon dots (Cdots) are a type of carbon nanoparticle that can be synthesized using various precursors and suitable methods, and they are employed in oil removal and recovery. A previous report illustrates the development of Cdots through a green methodology utilizing guava leaves and microwave irradiation (Fig. 6) [123]. The findings of this report highlight the effectiveness of carbon nanoparticles in oil recovery, which underscores the need for further exploration of such carbon-based materials in oil spill cleanup.

**Fig. 6** Motor oil removed from oil/water mixture by guava leaves derived carbon dots coated cotton. Reproduced from the reference [123]

## 3 Future Perspective

The ongoing development of sustainable carbon-based materials for oil removal and recovery continues to advance. While carbon-based materials are generally less hazardous than other composite materials, several recommendations for future research and development can be proposed. The foremost recommendation is the development of methods that can accurately determine the specific absorption capacity with a high degree of confidence. It is advisable to avoid reporting capacity within a broad range, which can be achieved by conducting a large number of repetitions, thereby enhancing recyclability.

Researchers should also focus on devising methods that are not only suitable for laboratory-scale experiments but also scalable for commercial or pilot-scale production. Bulk synthesis methods are essential to meet practical application demands. Additionally, researchers should strive to develop cost-effective synthesis methods with minimal mass loss. Therefore, alongside technical considerations, financial feasibility should be a significant concern in research endeavors.

Another critical challenge encountered by experts pertains to the regeneration of materials and the degradability of non-usable recovered oil. There is a need for improved regeneration capacity and the development of suitable approaches for the biological degradation of oil by microorganisms.

Future endeavours should aim to synthesize more efficient carbon-based sorbents with minimal, environmentally friendly, and practical modifications while preserving their desirable characteristics.

## 4 Conclusion and Outlook

Oil spills continue to pose significant threats to ecosystems in our current developmental era. This chapter has provided a concise overview of solutions aimed at mitigating the adverse impacts of oil–water mixtures, with a specific focus on oil removal and recovery. Effective sorbent materials must meet two essential criteria: a substantial surface area, porosity with capillary action, and hydrophobicity. Carbon-based materials excel in meeting these requirements, surpassing conventional alternatives.

This chapter has addressed the multifaceted challenges associated with oil contamination in water, delineating its environmental and aquatic life repercussions. It has explored a variety of carbon materials employed for oil spill remediation, encompassing their synthesis processes, properties, and characterization. Their biodegradability and chemical inertness make them more versatile than polymer-based materials.

Carbon-containing materials such as carbon nanotubes (CNTs), carbon nanofibers (CNFs), activated carbon, graphene and its derivatives, graphite, graphene aerogels, and carbon membranes have been introduced for their exceptional oil absorption capacity and remarkable reusability. They exhibit water-repellent and oleophilic properties while possessing robust mechanical, thermal, physical, and chemical characteristics. These materials can absorb oil at rates ranging from 3 to 913 times their own weight. Some are coated onto polyurethane (PU) foam and sponges to enhance recyclability and absorption capacity without significant deformation.

In conclusion, carbon-containing materials have proven to be efficient and effective in removing oil spills without causing harm to the ecosystem. Future research should continue to refine and expand upon these promising solutions to ensure the sustainable management of oil-contaminated water bodies.

## References

1. Piperopoulos E, Calabrese L, Mastronardo E, Milone C, Proverbio E (2020) Carbon-based sponges for oil spill recovery. In: Carbon nanomaterials for Agri-Food and environmental applications. Elsevier 155–75
2. Gupta S, Tai NH (2016) Carbon materials as oil sorbents: a review on the synthesis and performance. J Mater Chem A Mater 4(5):1550–1565
3. Ge J, Zhao H, Zhu H, Huang J, Shi L, Yu S (2016) Advanced sorbents for oil-spill cleanup: recent advances and future perspectives. Adv Mater 28(47):10459–10490
4. Jernelöv A (2010) How to defend against future oil spills. Nature 466(7303):182–183
5. Campagna C, Short FT, Polidoro BA, McManus R, Collette BB, Pilcher NJ et al (2011) Gulf of Mexico oil blowout increases risks to globally threatened species. Bioscience 61(5):393–397
6. Al-anzi BS, Siang OC (2017) Recent developments of carbon based nanomaterials and membranes for oily wastewater treatment. RSC Adv 7(34):20981–20994
7. Sankaranarayanan S, Lakshmi DS, Vivekanandhan S, Ngamcharussrivichai C (2021) Biocarbons as emerging and sustainable hydrophobic/oleophilic sorbent materials for oil/water separation. Sustain Mater Technol 28:e00268
8. Kajitvichyanukul P, Hung YT, Wang LK (2011) Membrane technologies for oil–water separation. In: Membrane and desalination technologies. Totowa, NJ: Humana Press, 639–68

9. Shi Z, Zhang W, Zhang F, Liu X, Wang D, Jin J et al (2013) Ultrafast separation of emulsified oil/water mixtures by ultrathin free-standing single-walled carbon nanotube network films. Adv Mater 25(17):2422–2427
10. Xie A, Cui J, Yang J, Chen Y, Dai J, Lang J et al (2019) Photo-Fenton self-cleaning membranes with robust flux recovery for an efficient oil/water emulsion separation. J Mater Chem A Mater. 7(14):8491–8502
11. Kostka JE, Prakash O, Overholt WA, Green SJ, Freyer G, Canion A et al (2011) Hydrocarbon-degrading bacteria and the bacterial community response in Gulf of Mexico beach sands impacted by the Deepwater horizon oil spill. Appl Environ Microbiol 77(22):7962–7974
12. Atlas RM, Hazen TC (2011) Oil Biodegradation and bioremediation: a tale of the two worst spills in U.S. history. Environ Sci Technol 45(16):6709–15
13. Margesin R, Schinner F (2001) Biodegradation and bioremediation of hydrocarbons in extreme environments. Appl Microbiol Biotechnol 56(5–6):650–663
14. da Rosa JJ, Rubio J (2005) The FF (flocculation–flotation) process. Miner Eng 18(7):701–707
15. Broje V, Keller AA (2006) Improved mechanical oil spill recovery using an optimized geometry for the skimmer surface. Environ Sci Technol 40(24):7914–7918
16. Buist I, Potter S, Nedwed T, Mullin J (2011) Herding surfactants to contract and thicken oil spills in pack ice for in situ burning. Cold Reg Sci Technol 67(1–2):3–23
17. Khalil MF, El-Boghdady I, Lotfy ER (2022) Oil-recovery performance of a sponge-covered drum skimmer. Alex Eng J 61(12):12653–12660
18. Kujawinski EB, Kido Soule MC, Valentine DL, Boysen AK, Longnecker K, Redmond MC (2011) Fate of dispersants associated with the Deepwater horizon oil spill. Environ Sci Technol 45(4):1298–1306
19. Feng L, Zhang Z, Mai Z, Ma Y, Liu B, Jiang L et al (2004) A Super-Hydrophobic and super-Oleophilic coating mesh film for the separation of oil and water. Angew Chem 116(15):2046–2048
20. Li H, Wang J, Yang L, Song Y (2008) Superoleophilic and Superhydrophobic inverse opals for oil sensors. Adv Funct Mater 18(20):3258–3264
21. Bayat A, Aghamiri SF, Moheb A, Vakili-Nezhaad GR (2005) Oil spill cleanup from sea water by sorbent materials. Chem Eng Technol 28(12):1525–1528
22. Annunciado TR, Sydenstricker THD, Amico SC (2005) Experimental investigation of various vegetable fibers as sorbent materials for oil spills. Mar Pollut Bull 50(11):1340–1346
23. Xie J, Meng W, Wu D, Zhang Z, Kong H (2012) Removal of organic pollutants by surfactant modified zeolite: comparison between ionizable phenolic compounds and non-ionizable organic compounds. J Hazard Mater 231–232:57–63
24. Thines RK, Mubarak NM, Nizamuddin S, Sahu JN, Abdullah EC, Ganesan P (2017) Application potential of carbon nanomaterials in water and wastewater treatment: a review. J Taiwan Inst Chem Eng 72:116–133
25. Chen B, Ma Q, Tan C, Lim T, Huang L, Zhang H (2015) Carbon-based sorbents with three-dimensional architectures for water remediation. Small 11(27):3319–3336
26. Nardecchia S, Carriazo D, Ferrer ML, Gutiérrez MC, del Monte F (2013) Three dimensional macroporous architectures and aerogels built of carbon nanotubes and/or graphene: synthesis and applications. Chem Soc Rev 42(2):794–830
27. Biener J, Stadermann M, Suss M, Worsley MA, Biener MM, Rose KA et al (2011) Advanced carbon aerogels for energy applications. Energy Environ Sci 4(3):656
28. Lu Y, Niu Z, Yuan W (2019) Multifunctional magnetic superhydrophobic carbonaceous aerogel with micro/nano-scale hierarchical structures for environmental remediation and energy storage. Appl Surf Sci 480:851–860
29. Mehmood A, Khan FSA, Mubarak NM, Mazari SA, Jatoi AS, Khalid M et al (2021) Carbon and polymer-based magnetic nanocomposites for oil-spill remediation—a comprehensive review. Environ Sci Pollut Res 28(39):54477–54496
30. Chen H, Müller MB, Gilmore KJ, Wallace GG, Li D (2008) Mechanically strong, electrically conductive, and biocompatible graphene paper. Adv Mater 20(18):3557–3561

31. Neuberger N, Adidharma H, Fan M (2018) Graphene: a review of applications in the petroleum industry. J Pet Sci Eng 167:152–159
32. Riaz MA, McKay G, Saleem J (2017) 3D graphene-based nanostructured materials as sorbents for cleaning oil spills and for the removal of dyes and miscellaneous pollutants present in water. Environ Sci Pollut Res 24(36):27731–27745
33. Rao CNR, Sood AK, Subrahmanyam KS, Govindaraj A (2009) Graphene: the new two-dimensional nanomaterial. Angew Chem Int Ed 48(42):7752–7777
34. Rao CNR, Sood AK, Voggu R, Subrahmanyam KS (2010) Some novel attributes of graphene. J Phys Chem Lett 1(2):572–580
35. Luo X, He Z, Gong H, He L (2022) Recent advances in oil-water separation materials with special wettability modified by graphene and its derivatives: a review. Chem Eng Process-Process Intensif 170:108678
36. Chen Z, Dong L, Yang D, Lu H (2013) Superhydrophobic graphene-based materials: surface construction and functional applications. Adv Mater 25(37):5352–5359
37. Nguyen DD, Tai NH, Lee SB, Kuo WS (2012) Superhydrophobic and superoleophilic properties of graphene-based sponges fabricated using a facile dip coating method. Energy Environ Sci 5(7):7908
38. Cong HP, Chen JF, Yu SH (2014) Graphene-based macroscopic assemblies and architectures: an emerging material system. Chem Soc Rev 43(21):7295–7325
39. Li D, Müller MB, Gilje S, Kaner RB, Wallace GG (2008) Processable aqueous dispersions of graphene nanosheets. Nat Nanotechnol 3(2):101–105
40. Kemp KC, Seema H, Saleh M, Le NH, Mahesh K, Chandra V et al (2013) Environmental applications using graphene composites: water remediation and gas adsorption. Nanoscale 5(8):3149
41. Ge B, Zhang Z, Zhu X, Men X, Zhou X, Xue Q (2014) A graphene coated cotton for oil/water separation. Compos Sci Technol 102:100–105
42. Hadji EM, Fu B, Abebe A, Bilal HM, Wang J (2020) Sponge-based materials for oil spill cleanups: a review. Front Chem Sci Eng 14(5):749–762
43. Sun H, Li A, Qin X, Zhu Z, Liang W, An J et al (2013) Three-dimensional superwetting mesh film based on graphene assembly for liquid transportation and selective absorption. Chemsuschem 6(12):2377–2381
44. Iqbal MZ, Abdala AA (2013) Oil spill cleanup using graphene. Environ Sci Pollut Res 20(5):3271–3279
45. Toyoda M, Moriya K, Aizawa J ichi, Konno H, Inagaki M (2000) Sorption and recovery of heavy oils by using exfoliated graphite Part I: Maximum sorption capacity. Desalination 128(3):205–11
46. Wang G, Wang B, Park J, Yang J, Shen X, Yao J (2009) Synthesis of enhanced hydrophilic and hydrophobic graphene oxide nanosheets by a solvothermal method. Carbon N Y 47(1):68–72
47. Periasamy AP, Wu WP, Ravindranath R, Roy P, Lin GL, Chang HT (2017) Polymer/reduced graphene oxide functionalized sponges as superabsorbents for oil removal and recovery. Mar Pollut Bull 114(2):888–895
48. Singhal AV, George R, Sharma AK, Malwal D, Lahiri I (2020) Development of superhydrophillic tannic acid-crosslinked graphene oxide membranes for efficient treatment of oil contaminated water with enhanced stability. Heliyon 6(10):e05127
49. Abdalla O, Wahab MA, Abdala A (2020) Mixed matrix membranes containing aspartic acid functionalized graphene oxide for enhanced oil-water emulsion separation. J Environ Chem Eng 8(5):104269
50. Liu Z, Wu W, Liu Y, Qin C, Meng M, Jiang Y et al (2018) A mussel inspired highly stable graphene oxide membrane for efficient oil-in-water emulsions separation. Sep Purif Technol 199:37–46
51. He Y, Liu Y, Wu T, Ma J, Wang X, Gong Q et al (2013) An environmentally friendly method for the fabrication of reduced graphene oxide foam with a super oil absorption capacity. J Hazard Mater 260:796–805

52. Niu Z, Chen J, Hng HH, Ma J, Chen X (2012) A leavening strategy to prepare reduced graphene oxide foams. Adv Mater 24(30):4144–4150
53. Iijima S (1991) Helical microtubules of graphitic carbon. Nature 354(6348):56–58
54. Zhu X, Zhang Z, Ren G, Yang J, Wang K, Xu X et al (2012) A novel superhydrophobic bulk material. J Mater Chem 22(38):20146
55. Pan B, Xing B (2008) Adsorption mechanisms of organic chemicals on carbon nanotubes. Environ Sci Technol 42(24):9005–9013
56. Mauter MS, Elimelech M (2008) Environmental applications of carbon-based nanomaterials. Environ Sci Technol 42(16):5843–5859
57. Maphutha S, Moothi K, Meyyappan M, Iyuke SE (2013) A carbon nanotube-infused polysulfone membrane with polyvinyl alcohol layer for treating oil-containing waste water. Sci Rep 3(1):1509
58. Chrzanowska J, Hoffman J, Małolepszy A, Mazurkiewicz M, Kowalewski TA, Szymanski Z, et al (2015) Synthesis of carbon nanotubes by the laser ablation method: effect of laser wavelength. Phys Status Solidi (b) 252(8):1860–7
59. Arora N, Sharma NN (2014) Arc discharge synthesis of carbon nanotubes: Comprehensive review. Diam Relat Mater 50:135–150
60. Annu A, Bhattacharya B, Singh PK, Shukla PK, Rhee HW (2017) Carbon nanotube using spray pyrolysis: recent scenario. J Alloys Compd 691:970–982
61. Zhou D, Anoshkina EV, Chow L, Chai G (2006) Synthesis of carbon nanotubes by electrochemical deposition at room temperature. Carbon N Y. 44(5):1013–1016
62. Shah KA, Tali BA (2016) Synthesis of carbon nanotubes by catalytic chemical vapour deposition: a review on carbon sources, catalysts and substrates. Mater Sci Semicond Process 41:67–82
63. Sun H, La P, Zhu Z, Liang W, Yang B, Zhao X et al (2014) Hydrophobic carbon nanotubes for removal of oils and organics from water. J Mater Sci 49(20):6855–6861
64. Ong CC, Sundera Murthe S, Mohamed NM, Perumal V, Mohamed Saheed MS (2018) Nanoscaled surface modification of poly(dimethylsiloxane) using carbon nanotubes for enhanced oil and organic solvent absorption. ACS Omega 3(11):15907–15915
65. Piperopoulos E, Calabrese L, Mastronardo E, Proverbio E, Milone C (2018) Synthesis of reusable silicone foam containing carbon nanotubes for oil spill remediation. J Appl Polym Sci 135(14)
66. Gui X, Zeng Z, Lin Z, Gan Q, Xiang R, Zhu Y et al (2013) Magnetic and highly recyclable macroporous carbon nanotubes for spilled oil sorption and separation. ACS Appl Mater Interfaces 5(12):5845–5850
67. Gui X, Wei J, Wang K, Cao A, Zhu H, Jia Y et al (2010) Carbon nanotube sponges. Adv Mater 22(5):617–621
68. Lee CH, Johnson N, Drelich J, Yap YK (2011) The performance of superhydrophobic and superoleophilic carbon nanotube meshes in water–oil filtration. Carbon N Y. 49(2):669–676
69. Ge B, Zhang Z, Zhu X, Ren G, Men X, Zhou X (2013) A magnetically superhydrophobic bulk material for oil removal. Colloids Surf A Physicochem Eng Asp 429:129–133
70. Liu J, Li X, Jia W, Ding M, Zhang Y, Ren S (2016) Separation of emulsified oil from oily wastewater by functionalized multiwalled carbon nanotubes. J Dispers Sci Technol 37(9):1294–1302
71. Wang H, Wang E, Liu Z, Gao D, Yuan R, Sun L et al (2015) A novel carbon nanotubes reinforced superhydrophobic and superoleophilic polyurethane sponge for selective oil–water separation through a chemical fabrication. J Mater Chem A Mater. 3(1):266–273
72. Gu J, Xiao P, Zhang L, Lu W, Zhang G, Huang Y et al (2016) Construction of superhydrophilic and under-water superoleophobic carbon-based membranes for water purification. RSC Adv 6(77):73399–73403
73. Zhang L, Gu J, Song L, Chen L, Huang Y, Zhang J et al (2016) Underwater superoleophobic carbon nanotubes/core–shell polystyrene@Au nanoparticles composite membrane for flow-through catalytic decomposition and oil/water separation. J Mater Chem A Mater. 4(28):10810–10815

74. Noamani S, Niroomand S, Rastgar M, Sadrzadeh M (2019) Carbon-based polymer nanocomposite membranes for oily wastewater treatment. NPJ Clean Water. 2(1):20
75. Zhu H, Qiu S, Jiang W, Wu D, Zhang C (2011) Evaluation of electrospun polyvinyl chloride/polystyrene fibers as sorbent materials for oil spill cleanup. Environ Sci Technol 45(10):4527–4531
76. Shang Y, Si Y, Raza A, Yang L, Mao X, Ding B et al (2012) An in situ polymerization approach for the synthesis of superhydrophobic and superoleophilic nanofibrous membranes for oil–water separation. Nanoscale 4(24):7847
77. Deschamps G, Caruel H, Borredon ME, Albasi C, Riba JP, Bonnin C, et al (2003) Oil removal from water by sorption on hydrophobic cotton fibers. 2. Study of sorption properties in dynamic mode. Environ Sci Technol 37(21):5034–9
78. Tang X, Si Y, Ge J, Ding B, Liu L, Zheng G et al (2013) In situ polymerized superhydrophobic and superoleophilic nanofibrous membranes for gravity driven oil–water separation. Nanoscale 5(23):11657
79. Baji A, Mai YW, Wong SC, Abtahi M, Chen P (2010) Electrospinning of polymer nanofibers: effects on oriented morphology, structures and tensile properties. Compos Sci Technol 70(5):703–718
80. Liu H, Cao CY, Wei FF, Huang PP, Sun YB, Jiang L et al (2014) Flexible macroporous carbon nanofiber film with high oil adsorption capacity. J Mater Chem A Mater. 2(10):3557
81. Tai MH, Gao P, Tan BYL, Sun DD, Leckie JO (2014) Highly efficient and flexible electrospun carbon-silica nanofibrous membrane for ultrafast gravity-driven oil-water separation. ACS Appl Mater Interfaces 6(12):9393–9401
82. Lin YZ, Zhong LB, Dou S, Shao ZD, Liu Q, Zheng YM (2019) Facile synthesis of electrospun carbon nanofiber/graphene oxide composite aerogels for high efficiency oils absorption. Environ Int 128:37–45
83. Visco A, Quattrocchi A, Nocita D, Montanini R, Pistone A (2021) Polyurethane foams loaded with carbon nanofibers for oil spill recovery: mechanical properties under fatigue conditions and selective absorption in oil/water mixtures. Nanomaterials 11(3):735
84. Ge X, Yang W, Wang J, Long D, Ling L, Qiao W (2015) Flexible carbon nanofiber sponges for highly efficient and recyclable oil absorption. RSC Adv 5(86):70025–70031
85. Xiao N, Zhou Y, Ling Z, Qiu J (2013) Synthesis of a carbon nanofiber/carbon foam composite from coal liquefaction residue for the separation of oil and water. Carbon N Y 59:530–536
86. Wu Z, Li C, Liang H, Chen J, Yu S (2013) Ultralight, flexible, and fire-resistant carbon nanofiber aerogels from bacterial cellulose. Angew Chem 125(10):2997–3001
87. Liang H, Guan Q, Chen L, Zhu Z, Zhang W, Yu S (2012) Macroscopic-scale template synthesis of robust carbonaceous nanofiber hydrogels and aerogels and their applications. Angew Chem Int Ed 51(21):5101–5105
88. Ieamviteevanich P, Palaporn D, Chanlek N, Poo-arporn Y, Mongkolthanaruk W, Eichhorn SJ et al (2020) Carbon nanofiber aerogel/magnetic core-shell nanoparticle composites as recyclable oil sorbents. ACS Appl Nano Mater 3(4):3939–3950
89. Guo Z, Long B, Gao S, Luo J, Wang L, Huang X et al (2021) Carbon nanofiber based superhydrophobic foam composite for high performance oil/water separation. J Hazard Mater 402:123838
90. Jin X, Al-Qatatsheh A, Subhani K, Salim NV (2021) Biomimetic and flexible 3D carbon nanofiber networks with fire-resistant and high oil-sorption capabilities. Chem Eng J 412:128635
91. Kim YA, Hayashi T, Endo M, Dresselhaus MS (2013) Carbon nanofibers. Springer handbook of nanomaterials, 233–62
92. Ng LY, Mohammad AW, Leo CP, Hilal N (2013) Polymeric membranes incorporated with metal/metal oxide nanoparticles: a comprehensive review. Desalination 308:15–33
93. Le NL, Duong PHH, Nunes SP (2017) 1.6 Advanced polymeric and organic–inorganic membranes for pressure-driven processes. In: Comprehensive membrane science and engineering. Elsevier 120–36

94. Li C, Song C, Tao P, Sun M, Pan Z, Wang T et al (2016) Enhanced separation performance of coal-based carbon membranes coupled with an electric field for oily wastewater treatment. Sep Purif Technol 168:47–56
95. Wu Y, Zhang X, Liu S, Zhang B, Lu Y, Wang T (2016) Preparation and applications of microfiltration carbon membranes for the purification of oily wastewater. Sep Sci Technol 51(11):1872–1880
96. Hsieh CT, Hsu JP, Hsu HH, Lin WH, Juang RS (2016) Hierarchical oil–water separation membrane using carbon fabrics decorated with carbon nanotubes. Surf Coat Technol 286:148–154
97. Liu Y, Shi Q, Hou C, Zhang Q, Li Y, Wang H (2017) Versatile mechanically strong and highly conductive chemically converted graphene aerogels. Carbon N Y. 125:352–359
98. Dong S, Xia L, Guo T, Zhang F, Cui L, Su X et al (2018) Controlled synthesis of flexible graphene aerogels macroscopic monolith as versatile agents for wastewater treatment. Appl Surf Sci 445:30–38
99. Ren RP, Li W, Lv YK (2017) A robust, superhydrophobic graphene aerogel as a recyclable sorbent for oils and organic solvents at various temperatures. J Colloid Interface Sci 500:63–68
100. Rodríguez-Mata V, González-Domínguez JM, Benito AM, Maser WK, García-Bordejé E (2019) Reduced graphene oxide aerogels with controlled continuous microchannels for environmental remediation. ACS Appl Nano Mater 2(3):1210–22
101. Shang Y, Yang H, Qin Z, Yin S, Yang L, Xu M et al (2020) Arbitrary-shaped reduced graphene oxide aerogels via an unsaturated water vapor reduction. Carbon N Y 168:169–179
102. Zhou J, Zhang Y, Yang Y, Chen Z, Jia G, Zhang L (2019) Silk fibroin-graphene oxide functionalized melamine sponge for efficient oil absorption and oil/water separation. Appl Surf Sci 497:143762
103. Liu C, Yang J, Tang Y, Yin L, Tang H, Li C (2015) Versatile fabrication of the magnetic polymer-based graphene foam and applications for oil–water separation. Colloids Surf A Physicochem Eng Asp 468:10–16
104. Shiu RF, Lee CL, Hsieh PY, Chen CS, Kang YY, Chin WC et al (2018) Superhydrophobic graphene-based sponge as a novel sorbent for crude oil removal under various environmental conditions. Chemosphere 207:110–117
105. Rahmani Z, Samadi MT, Kazemi A, Rashidi AM, Rahmani AR (2017) Nanoporous graphene and graphene oxide-coated polyurethane sponge as a highly efficient, superhydrophobic, and reusable oil spill absorbent. J Environ Chem Eng 5(5):5025–5032
106. Xia C, Li Y, Fei T, Gong W (2018) Facile one-pot synthesis of superhydrophobic reduced graphene oxide-coated polyurethane sponge at the presence of ethanol for oil-water separation. Chem Eng J 345:648–658
107. Pethsangave DA, Wadekar PH, Khose RV, Some S (2020) Super-hydrophobic carrageenan cross-linked graphene sponge for recovery of oil and organic solvent from their water mixtures. Polym Test 90:106743
108. Wang J, Shi Z, Fan J, Ge Y, Yin J, Hu G (2012) Self-assembly of graphene into three-dimensional structures promoted by natural phenolic acids. J Mater Chem 22(42):22459
109. Yuan J, Zhang M, Xia M, Cao W, Du M, Dou J et al (2019) Novel high-capacity and reusable carbonaceous sponges for efficient absorption and recovery of oil from water. Appl Surf Sci 487:398–408
110. Wei T, Fan Z, Luo G, Zheng C, Xie D (2009) A rapid and efficient method to prepare exfoliated graphite by microwave irradiation. Carbon N Y. 47(1):337–339
111. Bayat A, Aghamiri SF, Moheb A (2008) Oil sorption by synthesized exfoliated graphite (EG). Iran J Chem Eng 5(1):51–64
112. Hristea G, Budrugeac P (2008) Characterization of exfoliated graphite for heavy oil sorption. J Therm Anal Calorim 91(3):817–823
113. Toyoda M, Inagaki M (2000) Heavy oil sorption using exfoliated graphite. Carbon N Y. 38(2):199–210
114. Hu Y, Liu X, Zou J, Gu T, Chai W, Li H (2013) Graphite/Isobutylene-isoprene rubber highly porous Cryogels as new sorbents for oil spills and organic liquids. ACS Appl Mater Interfaces 5(16):7737–7742

115. Wang L, Fu X, Chang E, Wu H, Zhang K, Lei X et al (2014) Preparation and its adsorptive property of modified expanded graphite nanomaterials. J Chem 2014:1–5
116. Shokry H, Elkady M, Salama E (2020) Eco-friendly magnetic activated carbon nano-hybrid for facile oil spills separation. Sci Rep 10(1):10265
117. Anwana Abel U, Rhoda Habor G, Innocent OO (2020) Adsorption studies of oil spill clean-up using coconut coir activated carbon (CCAC). Am J Chem Eng 8(2):36
118. Raj KG, Joy PA (2015) Coconut shell based activated carbon–iron oxide magnetic nanocomposite for fast and efficient removal of oil spills. J Environ Chem Eng 3(3):2068–2075
119. Sun H, Li A, Zhu Z, Liang W, Zhao X, La P et al (2013) Superhydrophobic activated carbon-coated sponges for separation and absorption. Chemsuschem 6(6):1057–1062
120. Ani JU, Akpomie KG, Okoro UC, Aneke LE, Onukwuli OD, Ujam OT (2020) Potentials of activated carbon produced from biomass materials for sequestration of dyes, heavy metals, and crude oil components from aqueous environment. Appl Water Sci 30;10(2):69
121. Dai W, Kim SJ, Seong WK, Kim SH, Lee KR, Kim HY, et al (2013) Porous carbon nanoparticle networks with tunable absorbability. Sci Rep 28;3(1):2524
122. Shi H, Shi D, Yin L, Yang Z, Luan S, Gao J et al (2014) Ultrasonication assisted preparation of carbonaceous nanoparticles modified polyurethane foam with good conductivity and high oil absorption properties. Nanoscale 6(22):13748–13753
123. Varshney N, Tariq M, Arshad F, Sk MP (2022) Biomass-derived carbon dots for efficient cleanup of oil spills. J Water Process Eng. 49:103016

# Chitosan Based Materials for Oil Spill Treatment

**Nayeem Ahmed**

**Abstract** Oil spills on the sea surface cause irreparable damage to marine habitats and ecosystems, mainly caused by oil spills from tankers, offshore platforms, oil drilling platforms and oil wells. The use of adsorbents to absorb and recover oil spills is one of the most advantageous strategies for oil spill treatment because it is economical, has a high recovery rate, and does not cause secondary pollution of the ecosystem. Chitosan is the second most abundant biopolysaccharide in nature and is commonly extracted from crab and lobster shells. It is a biocompatible, biodegradable, non-toxic and renewable substance. It also has reactive amino and hydroxyl groups that can be easily modified to be hydrophobic for oil adsorption. Adsorbents made from chitosan have recently attracted great interest because they are cheap, abundant, biodegradable and ecological. This chapter will discuss some important chitosan-based materials which have been used for oil spill treatment and found to be very efficient.

**Keywords** Chitosan · Chitosan modified · Oil spill · Treatment · Biopolymer · Eco-friendly

## 1 Introduction

Chitosan is a linear polysaccharide derived from chitin, a natural polymer found in the exoskeletons of crustaceans, insects and fungi. It has been used for centuries for a variety of purposes, including as a wound dressing and natural insecticide. The chemical structure of chitosan is similar to that of cellulose, with the main difference being the presence of amino groups in the polymer chain. These amino groups impart unique properties to chitosan, including gel-forming ability, biocompatibility, and antimicrobial activity. Chitosan can be obtained from a variety of sources, including shrimp, crab, lobster and krill shells, as well as fungal sources and insect cuticles

N. Ahmed (✉)
Department of Chemistry, University of Kashmir, Srinagar, India
e-mail: Nayeem2329@gmail.com

© The Author(s), under exclusive license to Springer Nature Singapore Pte Ltd. 2025

S. A. Bhawani et al. (eds.), *Biobased Materials and Their Composites for Oil Spill Treatment*, Composites Science and Technology,
https://doi.org/10.1007/978-981-96-6049-0_8

[1]. Chitosan has a wide range of applications in various fields, including medicine, environmental science, and industry. In medicine, chitosan is used in wound healing, drug delivery, and tissue engineering due to its biocompatibility and ability to promote cell growth [2]. In environmental science, chitosan is used for water treatment, soil improvement, and as a component in biodegradable plastics. In industry, chitosan is used for food preservation, textile production, and paper manufacturing due to its antimicrobial properties and ability to improve material properties [3].

The production of chitosan involves the extraction of chitin from the source material, followed by a chemical deacetylation process to remove the acetyl groups and convert chitin into chitosan. Extraction methods can vary depending on the source material, with the most common being acid hydrolysis and enzymatic digestion. After extraction, chitosan must be purified and modified to achieve the desired properties for the intended application. Purification techniques include dialysis, ultrafiltration, and precipitation, while modification techniques include crosslinking, grafting, and blending [4]. Quality control measures must also be taken to ensure the safety and consistency of chitosan products, including testing for heavy metals and microbial contamination.

Oil spills have significant environmental and economic impacts and therefore require effective mitigation strategies. Environmental impacts depend on coastal type, environmental conditions and seasonal changes, while the fate of oil depends on coastal type, tidal energy and environmental conditions [5]. Effective monitoring data management is crucial for obtaining spatiotemporal data. This can be achieved through in-situ sensors and atmospheric data using remote sensing technology. Cleanup options include preventive methods such as physical, mechanical, chemical and biotechnological methods and oil recovery strategies (Fig. 1a). However, effective cleaning requires a detailed understanding of the morphology and wave currents of the coastal environment [6]. Seasonal fluctuations should also be considered when planning oil response strategies. Oil behavior changes under high temperatures, icy ground, and high wind conditions. Oil spills threaten marine ecosystems and coastlines, requiring the use of physical barriers, adsorbents, and dispersants to prevent oil from reaching the coast [7]. Oil spills may occur for a variety of reasons, such as pipeline rupture, tanker sinking, or drilling operation failure [8]. Large-scale oil spills can have long-lasting effects on economies and ecosystems, thus future efforts should concentrate on creating more potent mitigation techniques to lessen their effects. Comparing the years 2010–2019 to 2000–2009, there has been a decrease in medium- and large-scale spills of 71.7% and 43.75%, respectively. Nonetheless, there are still difficulties in mitigating and remediating oil spills, with the northeastern United States, the Gulf of Mexico, and the Mediterranean Sea being the primary hotspots [5]. Therefore, it is crucial to keep up efforts to develop efficient means for responding to oil spills, such as physical, chemical, and biological remediation procedures, and to track and map oil spills using a variety of monitoring techniques, such as remote sensing and GIS tools.

A natural biopolymer derived from chitin called chitosan has demonstrated remarkable promise as a biosorbent for removing oil from emulsions of crude oil and salt water. Because of its widespread availability, biodegradability, and low toxicity,

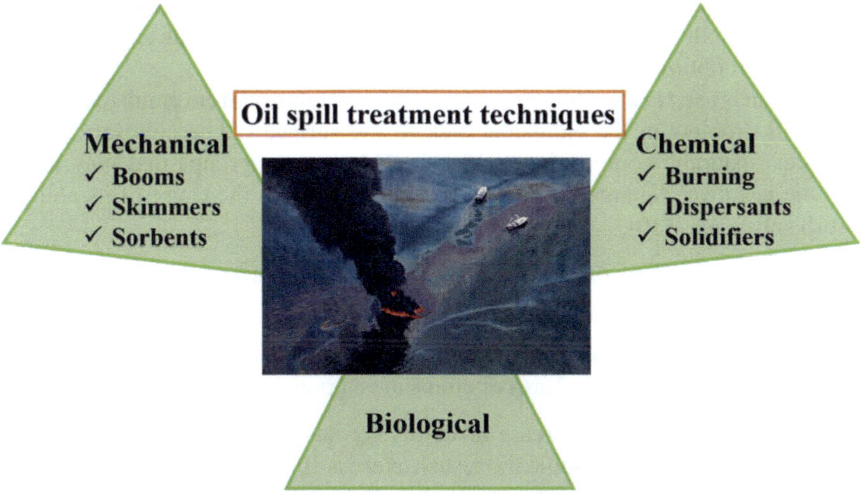


**a:** Common techniques used in oil spill treatment (Reproduced with permission).

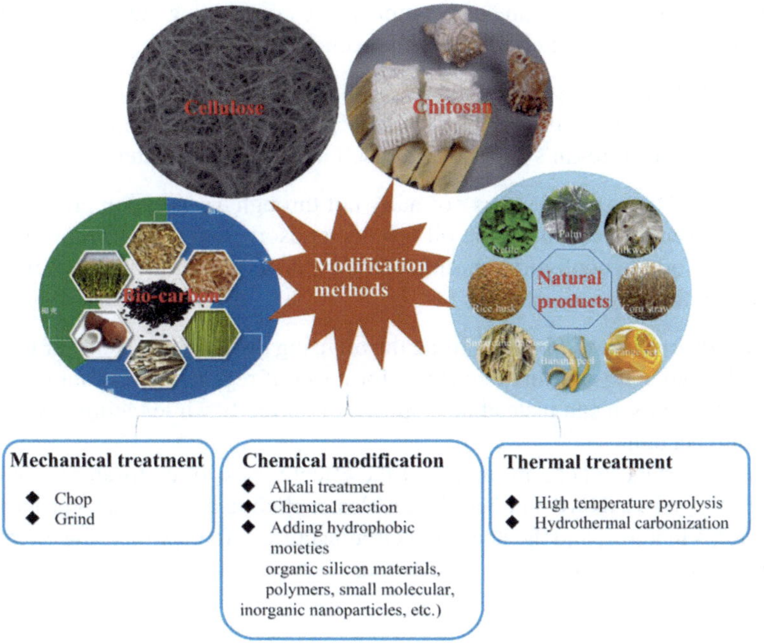


**b:** Modification methods for biomass materials. (Reproduced with permission).

**Fig. 1** **a** Common techniques used in oil spill treatment (Reproduced with permission). **b** Modification methods for biomass materials (Reproduced with permission)

chitosan can be used as an efficient biosorbent [9]. It is also very versatile in the sense that various modifications, through different methods, can be made and applications tailored for it optimum activity and performance (Fig. 1b).

Chitosan is used in the adsorption stage of the oil removal from emulsions process, where oil droplets in the emulsion stick to the surface of the chitosan particles. Chitosan's cationic composition, which includes amino groups that may interact with the negatively charged oil droplets, is thought to be the cause of its ability to absorb substances [10].

*The General Steps for Utilizing Chitosan to Remove Oil Are as Follows*

*Chitosan Preparation*: Chitin, a substance that is often found in the shells of crustaceans like shrimp or crab, may be purchased commercially or produced to make chitosan. The most common form of chitosan is a powder.

*Activation of chitosan*: To increase its ability for adsorption, chitosan is often activated. Through chemical modification, such as cross-linking or functionalization with other chemicals, activation can be accomplished. In this stage, chitosan's surface area and reactivity are increased.

*Preparation of the emulsion*: Prior to treatment, the crude oil-in-saline water emulsion needs to be created. This can be accomplished by combining the crude oil with salty water and stabilising the emulsion with mechanical agitation or surfactants.

*Treatment of the chitosan*: To enable adsorption, the chitosan is added to the emulsion and well mixed. When the oil droplets and chitosan particles come into contact, the oil sticks to the chitosan's surface as a result of electrostatic interactions.

*Separation*: The combination is normally put through a separation procedure to get rid of the treated water after the oil has been adsorbed onto the chitosan particles. Depending on the configuration, this could be done via centrifugation, filtering, or another method.

*Reusability or disposal*: By desorbing the oil using the proper solvents or treatments, the oil-loaded chitosan can be renewed for reuse. If regeneration is not possible, the oil-loaded chitosan can instead be disposed of correctly while taking environmental rules into account.

It is important to keep in mind that the effectiveness of chitosan as a biosorbent might change based on a number of variables, including the chitosan's properties, the emulsion's features, and the operating circumstances. To maximise the effectiveness of oil removal, it is crucial to optimise these parameters.

## 1.1 Chitosan-g-Citronellal Schiff Base Polymer

An efficient way to remove oil spills from water surfaces is to use a new amphiphilic Chitosan-g-Citronellal Schiff Base Polymer [11]. In comparison to native chitosan, the polymer with a hydrophobic citronellal grafted moiety (Fig. 2) demonstrated an oil adsorption capacity enhancement of 166% and 120% for light and heavy crude oil, respectively. The Ch-Cit schiff base's amphiphilic nature was regulated by altering the citronellal grafting percentage (G%) between 11 and 61%. Additionally, the Ch-Cit schiff base demonstrated greater hydrophobicity than native chitosan, and the hydrophobic properties of the synthesised schiff base derivatives were further increased by the addition of citronellal to the feed mixture. When the schiff base was forming, chemical crosslinking with glutaraldehyde produced holes and decreased the hydrophilicity of the chitosan backbone. The surface area of the synthesised Ch-Cit schiff base was greater (115.94 $m^2/g$) than that of plain chitosan (57.78 $m^2/g$) [1]. Additionally, because of its enhanced hydrophobic nature, the chitosan-g-citronellal schiff base polymer demonstrated a greater capacity for oil absorption than native chitosan. Response surface methodology (RSM) was also used in the study to identify the ideal operating parameters for the adsorptive cleanup of oil spills. After examining the impact of grafting percentage on the absorption capacity of various oils, it was discovered that adsorbent grafting percentages enhanced oil capacity values exponentially. The Chitosan-g-Citronellal Schiff Base Polymer's adsorption ability was assessed in terms of its affinity for the adsorptive removal of various oil types under various operational adsorption settings. For additional adsorption research, light and heavy crude oils were used, and the polymer demonstrated suitable adsorption abilities towards crude oils. According to the study, the agitation speed, adsorbent dose, and oil amount all have an impact on the polymer's ability to absorb oil. The overall goal of the project is to remove oil spills as effectively and economically as possible by employing a novel amphiphilic Chitosan-g-Citronellal Schiff Base Polymer (Table 1).

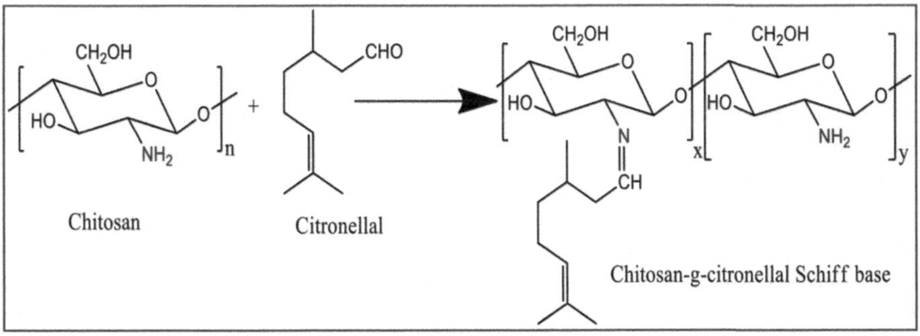


**Fig. 2** Proposed mechanism for the synthesis of chitosan-g-citronellal schiff base (Ch-Cit) (Reproduced with permission)

**Table 1** Comparison of maximum sorption capacity for the crude oil by various sorbents (Reproduced with permission)

| Adsorbent | Sorption capacity |
|---|---|
| Chitosan flakes | 0.38 g/g |
| Chitosan powder | 0.28 g/g |
| Chitosan-based polyacrylamide hydrogel | 2.31 g/g |
| Chitosan (prawn shells) | 18.52 g/g |
| Acetylated corncobs | 0.08 mg/g |
| Lauric acid (LA) modified oil palm leaves | 1.2 mg/g |
| Carbonized rice husks (CRH) | 6 g/g |
| Cellulose aerogel functionalized with methyltrimethoxysilane | 24.41 g/g |
| Butyl rubber | 25 g/g |
| Non-woven polypropylene | 15 g/g |
| BuAc cellusoe graft copolymer | 13.83 g/g |
| Chitosan-g-citronellal schiff base | 24.40 g/g |

## *1.2 Chitosan-g-Octanal Schiff Base Polymer*

The purpose of the study was to evaluate the effectiveness of a new amphiphilic chitosan-g-octanal Schiff base polymer in oil spill cleanup [12]. By altering the citronellal grafting %, the polymer's amphiphilic nature could be adjusted. This resulted to modifications in the polymer's ion exchange capacity, solubility, water absorption profiles, and oil adsorption capacity (Fig. 3). FT-IR spectroscopy and a $^{1}H$ NMR spectrometer were used to confirm the polymer's chemical structure, while SEM and BET analysis were used to examine morphological alterations and surface area [12]. In comparison to neat chitosan, the synthesised polymer had a larger surface area and a noticeably greater ability to adsorb both light and heavy crude oils. The hydrocarbon long chains of citronellal and the chitosan's decreased hydrophilic amine groups improved the oil sorption absorption. The oil capacity values rose exponentially with increasing polymer grafting percentages. The utilisation of response surface methodology (RSM) was employed to enhance many operational parameters, including agitation speed, oil amount, adsorption time, and adsorbent dose. Because of its increased hydrophobicity following Schiff base formation, the chitosan-g-octanal Schiff base polymer demonstrated a considerably greater oil adsorption capacity as compared to chitosan alone. Under a variety of seawater environmental circumstances, including oil kind, oil volume, adsorbent dose, and contact time, the polymer's effectiveness in removing oil spills was assessed. All things considered, the investigation proved that the new amphiphilic chitosan-g-octanal Schiff base polymer may be used as a powerful adsorbent to remove oil spills (Fig. 4).

The study's findings demonstrated that Chitosan-*g*-Octanal Schiff base had a much larger capacity for oil adsorption than Chitosan, likely as a result of an increase in its

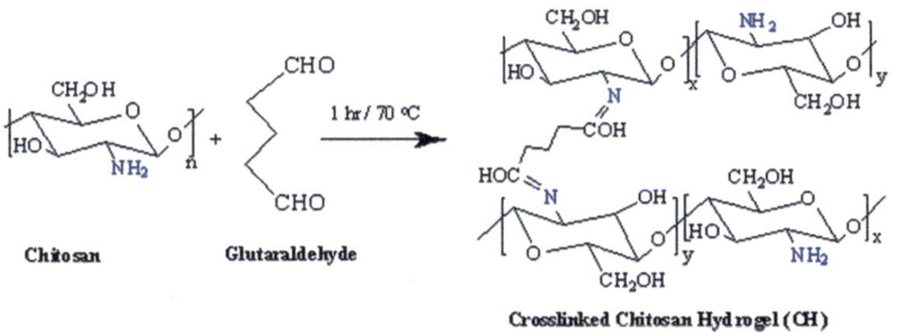


(A) Chitosan

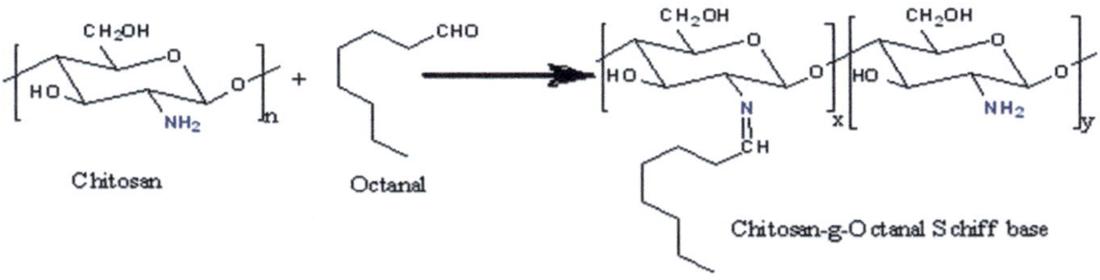


(B) Chitosan-g-Octanal Schiff base

**Fig. 3** **a** Chitosan and **b** Chitosan-g-Octanal Schiff base adsorbents formation (Reproduced with permission)

hydrophobic characteristics during Schiff base synthesis. Positively, the adsorption capacities of mineral oil, kerosene, diesel, light crude oil, and heavy crude oil all increase with the preparation of adsorbents. The investigation also showed that, at a constant contact time of 180 min, adsorbent amount (0.1 g), oil amount (0.5 g), shaking rate (150 rpm), and adsorption temperature medium (25 °C), the maximum oil adsorption percentage (%) for Chitosan-g-Octanal Schiff base was 96.41%, while it was 64.99% for native Chitosan.

## *1.3 Chitosan-Based Polyacrylamide Hydrogels*

Many studies have been conducted on the adsorption of crude oil from aqueous solutions utilising polyacrylamide hydrogels based on chitosan that were made by radiation-induced graft polymerization (Fig. 5) [13]. After conducting equilibrium studies to ascertain the hydrogel's adsorption capacity, it was discovered that the hydrogel produced with a 40% acrylamide concentration and a 5 kGy radiation dose demonstrated a high crude oil removal efficiency. A pseudo-second-order kinetic model was established to govern the adsorption of oil onto the hydrogel, suggesting

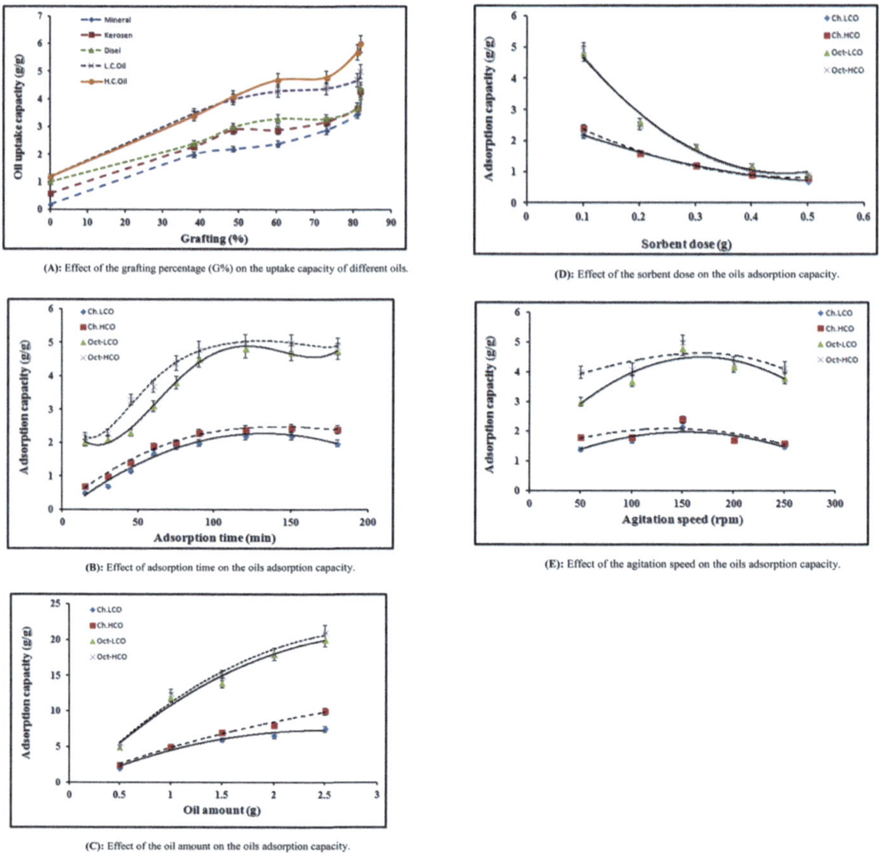


**Fig. 4** Impact of grafting percentage (G%) on various oils' absorption capacity **a**, impact of adsorption time **b**, volume of oil **c**, dose of sorbent **d**, and speed of agitation **e** on the oil's adsorption capacity (Reproduced with permission)

a more intricate adsorption mechanism than the pseudo-first-order model. The equilibrium data from the trials fit the Freundlich and Langmuir models quite well, indicating the presence of several adsorption mechanisms. To find the ideal adsorption conditions, experimental factors including the beginning concentrations of oil residue, acrylamide concentration, contact time, and pH were changed. In addition, FTIR and SEM micrographs were used to characterise the hydrogel in order to comprehend its morphology and structure. These results advance our knowledge of the adsorption behaviour of polyacrylamide hydrogels based on chitosan and offer information about possible uses for the materials in the extraction of crude oil from aqueous solutions.

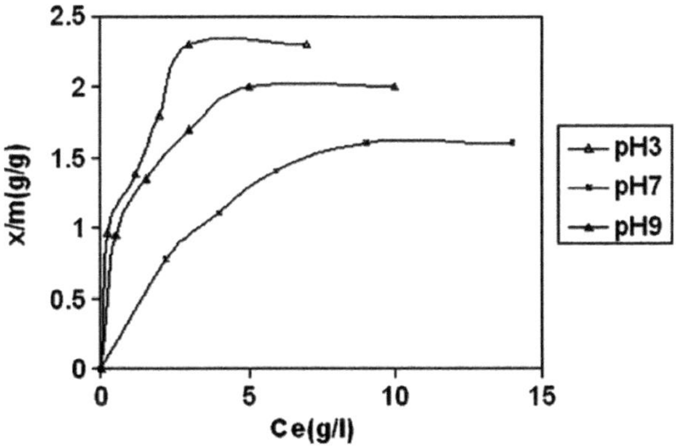


**Fig. 5** Equilibrium isotherm of the adsorption of crude oil on to the prepared hydro- gel (40% AM at radiation dose 5 kGy) (Reproduced with permission)

## *1.4 Chitosan Derived from Seafood Industry Waste for the Purpose of Oil Spill Cleanup.*

Chitosan, which is produced from leftovers from the seafood industry like prawn shells, has demonstrated significant promise as an efficient and environmentally friendly substitute for cleaning up oil spills [14]. The adsorption capabilities of chitosan are responsible for its positive charge, which enables it to efficiently extract oil from oil–water mixtures. To understand their effects on oil removal using chitosan, a number of parameters have been examined, including contact time, pH, starting concentration, and mass of adsorbent. It has been discovered that oil can be removed more effectively in an acidic media, demonstrating the significance of pH in the adsorption process. Pseudo-second order model-based adsorption of chitosan has been used in batch studies to investigate the kinetics of oil removal. The Langmuir model was used to fit the equilibrium data, revealing a maximum adsorption capacity of 30.30 g/g at 25 °C. Furthermore, a variety of methods, including TGA, SEM, and FT-IR spectroscopy, have been used to characterise chitosan and chitosan-g-Octanal Schiff base adsorbents. Because octanal is hydrophobic and chitosan's amine groups are consumed, the amphiphilic Chitosan-g-Octanal Schiff base adsorbent has been discovered to be more hydrophobic than chitosan. This increased hydrophobicity enhances its affinity towards oils and decreases its affinity towards water. Overall, chitosan has been proven to have high adsorption capacity and can be used as an effective material for removing oil spills (Fig. 6).

## *1.5 Chitosan as a Biosorbent for Oil Removal from Crude Oil-in-Saline Water Emulsions*

Chitosan, a biomaterial derived from chitin, has shown great potential as an efficient biosorbent for oil removal from crude oil-in-saline water emulsions (Fig. 7). The

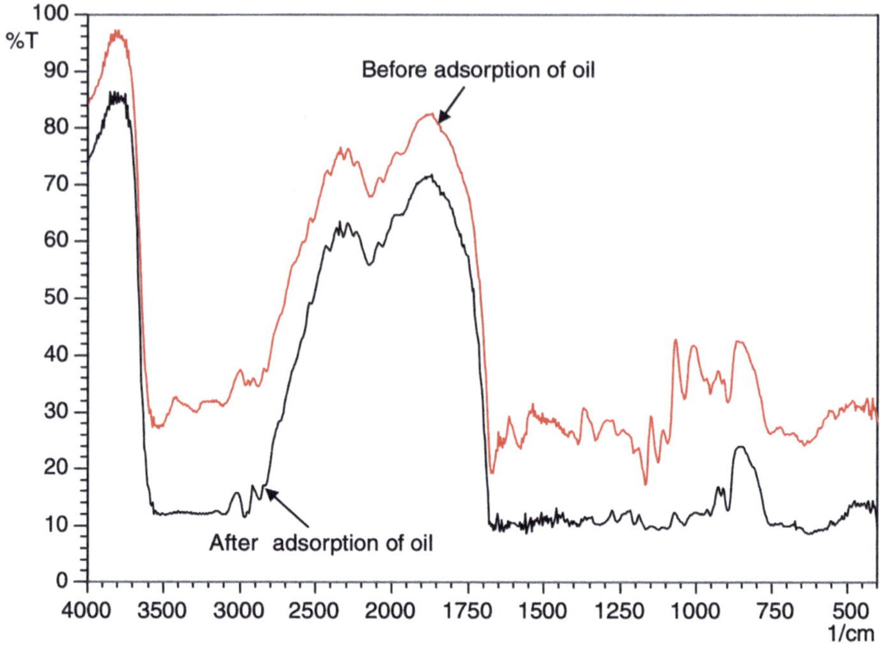


**Fig. 6** FT-IR spectra of chitosan before and after adsorption of oil (Reproduced with permission)

adsorption process of chitosan is influenced by various factors, including pH levels. In acidic conditions, the adsorption process is catalyzed, allowing oil droplets to fit into the active sites on the chitosan molecule. Chitosan has been found to have a high oil absorption capacity, with the ability to absorb up to 99.9 mg/g of oil [15]. Additionally, studies have shown that chitosan present in the cell walls of fungal biomass, such as Mucor rouxii, contributes to their oil adsorption capacity of up to 90%. The efficiency of chitosan for oil removal is influenced by parameters such as pH, moisture content, porosity, and degree of de-acetylation. Lower pH levels have been found to result in higher oil removal rates, while an increase in pH decreases the oil removal rate. Chitosan with a pH of 6.8, moisture content of 2.8%, porosity of 60–65%, and degree of de-acetylation higher than 85% has been found to be effective for oil removal. Furthermore, chitosan can be chemically and physically modified to enhance its solubility in water. Modified forms of chitosan have been used in various industries for contaminant removal from water. Overall, chitosan's unique properties and ability to effectively remove oil from crude oil-in-saline water emulsions make it a promising biosorbent for oil spills and pollution control applications (Table 2).

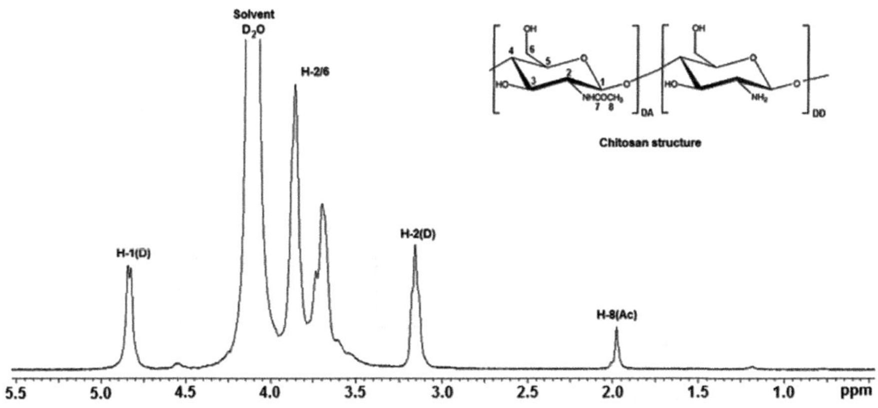


**Fig. 7** $^1$H NMR spectrum of chitosan (Reproduced with permission)

**Table 2** Oil removal from oil-in-water emulsions by chitosan reported in literature (Reproduced with permission)

| DD[a] of chitosan (%) | Oil studied | Dosage (g $L^{-1}$) | IOC[b] (mg $L^{-1}$) | pH | Contact time at equilibrium | Oil removal (%) |
|---|---|---|---|---|---|---|
| **90** | POME[c] | 0.5 | 10,000 | 5.0 | 15 min | 99 |
| **> 85** | SMO[e] | 0.2 | 200 | 5.0 | 6 h | 99 |
| | Canola oil | | | 7.6 | | 96 |
| | | | | 5.0 | | 99 |
| | | | | 7.6 | | 99 |
| | Cutting oil | | | 5.0 | | 97 |
| | | | | 7.6 | | 89 |
| **85** | Cutting oil | 8.0 | 1140 | 3.0 | 40 min | 45 |
| **94** | Heavy Crude oil | Not given | 200 | Not given | Not given | 90 |
| **86** | Crude oil | 0.5 | 30,000 | 3.0 | | 80 (pure water) |
| | | | | | | 60 (sea water) |
| | | | | | | 95 (saline water) |
| **93** | Light crude oil | 0.5 | 80 and 90 | 6.6 | 20 h | |

## 1.6 Dual Chemical Reactivity of Chitosan Based Materials for Oil Separation

Oil/water separation membranes have been made by optimizing super hydrophobicity and underwater superoleophobicity by the use of dual chemical reactivity at a single interface. A persistent superhydrophobic and submerged super-oleophobic material were produced by altering the biodegradable and naturally occurring chitosan (CTS) polymer by optimising the appropriate coexistence of critical topography and chemistry that contributed to extreme water and oil wettability [16]. By adding ethanol dropwise to the acidic (0.1 M acetic acid in water) chitosan solution, a granular nanoprecipitate of the polymer was created. This nanoprecipitate was then covalently cross-linked with 5-Acl in an alkaline environment using a 1,4-conjugate addition procedure.. Ultimately, this polymeric material generated from chitosan provided incredibly porous, hierarchically featured polymeric materials that could be molded into any stable shape that was needed. Water wettability tests were performed on the CTS gel material, which was naturally embedded with superhydrophilicity and had a water-beading contact angle of 0 degrees (Fig. 8).

The application of modified SSMs in oil/water separation has shown promise for universal applicability, exhibiting great efficiency with an impressive separation efficiency of over 98% and a water permeate flux of more than $2.6 \times 104\ \mathrm{L\ m^{-2}\ h^{-1}}$. Hydrophobic modification using a low-surface energy material can lead to superhydrophobicity, and hierarchical SSMs can have superhydrophobicity. Through the triggered change in dominance of two types of orthogonal functional groups in the material, this study successfully demonstrated gravity-driven filtration of desired liquids from oil/water mixtures using a single, multifunctional interface derived from a biodegradable polymer. The interface is capable of displaying two distinct durable biomimetic (fish scale/lotus leaf) liquid (oil/water) wettabilities. Even under a range of challenging, practically significant conditions, the CTS gel substance was incredibly effective in the failed separation of many types of oils, regardless of the density, viscosity, and surface tension of the liquids. The current design of biomimicked interfaces with two completely different orthogonal chemical functionalities has a great deal of potential for both fundamental research and a variety of applications.

## 1.7 Chitosan-Based Aerogels

Through various chemical changes, hydrophobic/oleophilic chitosan-based aerogels have been successfully generated and used for cleaning up oil spills [17]. Their oil absorption capacity, recycling ability, and oil–water selectivity are, nevertheless, significantly influenced by the fabrication techniques, hydrophobic modification agent, and micro-nano structure (Fig. 9). Zhang et al. produced a porous core/shell structure aerogel with high porosity (>98.16%) and low density (10.19–36.05 mg $\mathrm{cm}^3$), using CS aerogel as the core and CS hydrogel as the surface coating [18].

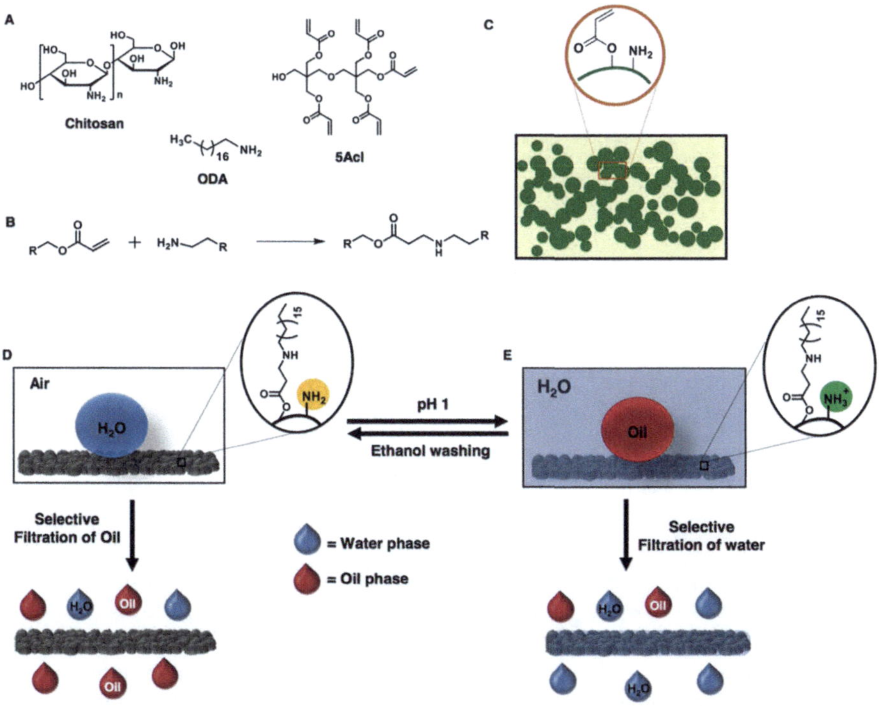


**Fig. 8** Design of single interface with switchable two distinct biomimicked wettability, **a** Structures of dipentaerythritol penta-acrylate (5Acl), octadecylamine (ODA) and chitosan (CTS). **b** demonstrates how the amine groups of chitosan and the acrylate groups of dipentaerythritol penta-acrylate react to generate an amphiphilic gel. **c** Shows that principal amine groups and leftover acrylate are present in the reaction product of 5Acl with chitosan. **d, e** The synthesized polymeric interface's pH-responsive, reversible switching between superhydrophobicity **d** and underwater superoleophobicity **e** properties through the protonation and deprotonation of primary amine groups; these interfaces were then used to selectively separate the aqueous or oily phases from various types of oil/water mixtures

Cao et al. designed and constructed a chitosan aerogel with oriented wave-shaped layer microstructures that shows superelasticity and hydrophobicity [19]. They used reduced graphene oxide and chitosan to make a composite aerogel that has superamphiphilicity in the air and superoleophobicity underwater. A new aerogel with exceptional mechanical stability, salt tolerance, and superoleophobicity was created by incorporating nanofibrillated cellulose into chitosan matrix.

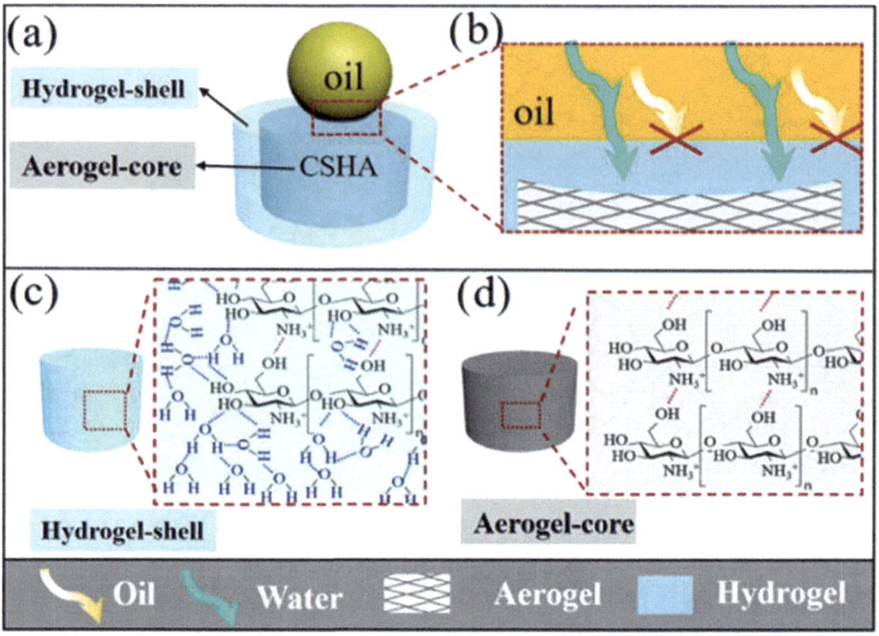


**Fig. 9** Mechanism of water selectively collected in CSHA: **a** The CSHA's core/shell-like structure **b** Water molecules can still pass through the aerogel core of the CSHA until all of the pores are saturated, but the hydrogel shell prevents additional oil molecule absorption. **c** Numerous water molecules absorbed on the CSHA's surface, along with the formation of hydrogen bonds within and between molecules (dashed lines in red and blue). The hydrogel and the CS aerogel are translated simultaneously. **d** Aerogel core schematic showing hydrogen bonds—red dashed lines—as the main type of intermolecular contact. (Reproduced with permission)

## 2 Conclusion

The creation of sorbent materials for the removal of organic contamination from water has a great deal of promise. Chitosan is a versatile material that may be utilised directly in a wide range of applications or used to create a wide range of materials, including hydro-gel beads, nanoparticles, films, membranes, and meshes. The ability of chitosan and products based on this polymer to effectively remove hydrocarbons from polluted water has led to its widespread use in cleaning oil spills. Numerous modification techniques have been widely used and effectively implemented to further increase the hydrophobicity and oil adsorption capability. Moreover, environmentally friendly and ecologically sustainable cleanup techniques are urgently required due to the serious harm that the release of spilled oil causes to the life of biological ecosystems and biodiversity. Without a doubt, because biomass-based sorbent materials are inexpensive, recyclable, and environmentally acceptable, they have shown some promise in the extraction and recovery of spilled oils. Thus, the primary area of research in the future will continue to be the creation of innovative chitosan-based oil sorbents that are effective, affordable, sustainable, intelligent, and friendly to the environment.

## References


1. (a) Liang XX, Omer AM, Hu Z, Wang Y, Yu D, Ouyang X (2018) Efficient adsorption of diclofenac sodium from aqueous solutions using magnetic amine-functionalized chitosan. Chemosphere 217;270–278; (b) Eskhan A, Banat F, Abu Haija M, Al-Asheh S (2019) Synthesis of mesoporous/macroporous microparticles using three dimensional assembly of chitosan-functionalized halloysite nanotubes and their performance in the adsorptive removal of oil droplets from water. Langmuir 35:2343–2357
2. (a) Mourya VK, Inamdar NN (2008) Chitosan-modifications and applications: opportunities galore. React Funct Polym 68:1013–1051; (b) Mohy Eldin MS, Ammar YA, Tamer TM, Omer AM, Ali AA (2016) Development of oleophilic adsorbent based on chitosan- poly (butyl acrylate) graft copolymer for petroleum oil spill removal. Int J Adv Res 4:2095–2111
3. Benner SW, John VT, Hall CK (2015) Simulation study of hydrophobically modified chitosan as an oil dispersant additive. J Phys Chem B 119:6979–6990
4. Tamer TM, Valachová K, Hassan MA, Omer AM, El-Shafeey M, Mohy Eldin MS, Šoltés L (2018) Chitosan/hyaluronan/edaravone membranes for anti-inflammatory wound dressing: in vitro and in vivo evaluation studies. Mater Sci Eng, C 90:227–235
5. Jenkins DW, Hudson SM (2001) Review of vinyl graft copolymerization featuring recent advances toward controlled radical-based reactions and illustrated with chitin/chitosan, Trunk Polymers. Chem Rev 101:3245–3274; (b) Nithya R, Sudha PN (2014) Grafting of N-butyl acrylate on to chitosan by ceric Ion initiation and its antimicrobial activity. Der Pharm Lett 6:58–64
6. Mohamed AH, Ahmed MO, Eman A, Walid MAB, Tamer MT (2018) Preparation, physico-chemical characterization and antimicrobial activities of novel two phenolic chitosan Schiff base derivatives. Sci Rep 8:1–14
7. (a) Srinivasan A, Viraraghavan T (2008) Removal of oil by walnut shell Media Bioresour. Technol 99:8217–8220 (b) Zunaira A, Chen Z, An C, Dong J (2022) Environmental impacts and challenges associated with oil spills on shorelines. J Mar Sci Eng 10:762
8. Raymon A, Karthik R (2015) Reclaiming aged transformer oil with activated bentonite and enhancing reclaimed and fresh transformer oils with antioxidants. IEEE Trans Dielectrics Elec Insulation 21:548–555
9. Ahmad AL, Sumathi S, Hameed BH (2005) Adsorption of residue oil from palm oil mill effluent using powder and flake chitosan: equilibrium and kinetic studies. Water Res 39:2483–2494
10. Ahmad AL, Sumathi S, Hameed BH (2005) Residual oil and suspended solid removal using natural adsorbents chitosan, bentonite and activated carbon: a comparative study. Chem Eng J 108:179–185
11. Omer AM, Eweida BY, Tamer TM et al (2021) Removal of oil spills by novel developed amphiphilic chitosan-g-citronellal schiff base polymer. Sci Rep 11:19879
12. Tamer TM, Eweida BY, Omer AM, Soliman HMA, Ali SM, Zaatot AA, Mohy-Eldin MS (2021) Removal of oil spills by novel amphiphilic Chitosan-g-Octanal Schiff base polymer developed by click grafting technique. J Saudi Chem Soc 25:101369
13. HH Sokker, El-Sawy NM, Hassan MA, El-Anadouli BE (2011) Adsorption of crude oil from aqueous solution by hydrogel of chitosan based polyacrylamide prepared by radiation induced graft polymerization. J Hazard Mater 190, 359–365
14. (a) Ummadisingu A, Gupta S (2012) Characteristics and kinetic study of chitosan prepared from seafood industry waste for oil spills cleanup. Desalination Water Treat 44(1–3):44–51; (b) Eweida BY, Omer AM, Tamer TM, et al (2023) Kinetics, isotherms and thermodynamics

of oil spills removal by novel amphiphilic Chitosan-g-Octanal Schiff base polymer developed by click grafting technique. Polym Bull 80:4813–4840
15. Vidal RRL, Desbrières J, Borsali R, Guibal E (2020) Oil removal from crude oil-in-saline water emulsions using chitosan as biosorbent. Sep Sci Technol 55(5):835–847
16. Jana N, Parbat D, Manna U (2019) Rational use of dual chemical reactivity in a single interface for optimizing both superhydrophobicity and underwater superoleophobicity. Chem Mater 31(5):1479–1484
17. Ouyang D, Lei X, Zheng H (2023) Recent advances in biomass-based materials for oil spill cleanup. Nanomaterials 13:620
18. Zhang E, Li W, Gao Y, Lei C, Huang H, Yang J, Zhang H, Li D (2020) High-capacity reusable chitosan absorbent with a hydrogel-coated/aerogel-Core structure and superhydrophilicity under oil for water removal from oil. ACS Appl Bio Mater 3:5872–5879
19. Cao M, Li SL, Cheng JB, Zhang AN, Wang YZ, Zhao HB (2021) Fully bio-based, low fire-hazard and superelastic aerogel without hazardous cross-linkers for excellent thermal insulation and oil clean-up absorption. J Hazard Mater 403:123977

# Chitosan-Silica Hybrid Nanomaterials for Oil Spill Treatment

**Alimuddin, Salman A. Khan, Abdul Moheman, Mohd. Amil Usmani, and Showkat Ahmad Bhawani**

**Abstract** Oil spills pose enormous environmental risks, affecting marine ecosystems, wildlife, and coastal towns. In the field of oil spill treatment, chitosan-silica hybrid nanomaterials are a unique approach that has several advantages over conventional techniques. They effectively adsorb oil and other hydrophobic substances from water due to their high surface area and porosity. This adsorption process is reversible; therefore, chitosan-silica hybrid nanomaterials can be recovered and reused. Furthermore, the biodegradable nature of chitosan assures that the adverse effect on the environment is minimized during the clean-up operation. Furthermore, chitosan-silica hybrid nanomaterials can be altered to improve their suitability for oil spill treatments. Studies have shown that chitosan-silica hybrid nanomaterials have significant potential as efficient and recyclable oil absorbents.

## 1 Introduction

Crude oil is a vital natural resource that enhances human and societal development. The development of the economy and society requires massive oil extraction and sea transport. However, frequent oil spills and leaks generate significant economic losses, harm marine ecosystems, and may endanger human life [1]. Oils are a major source of water contamination worldwide [2]. Consequently, oil spill cleanup has received widespread scientific and industrial attention.

Alimuddin · S. A. Khan · A. Moheman (✉)
Department of Chemistry, School of Sciences, Maulana Azad National Urdu University, Hyderabad 500032, Telangana, India
e-mail: amohemanappchem@gmail.com

Mohd. A. Usmani
Department of Chemistry, Gandhi Faiz-e-Aam College, Shahjahanpur (Affiliated to Mahatma Jyotiba Phule Rohilkhand University, Bareilly), Shahjahanpur 242001, Uttar Pradesh, India

S. A. Bhawani
Faculty of Resource Science and Technology, Universiti Malaysia Sarawak (UNIMAS), Kuching 94300, Sarawak, Malaysia

© The Author(s), under exclusive license to Springer Nature Singapore Pte Ltd. 2025

S. A. Bhawani et al. (eds.), *Biobased Materials and Their Composites for Oil Spill Treatment*, Composites Science and Technology,
https://doi.org/10.1007/978-981-96-6049-0_9

Numerous methods have been developed to clean and recover oil spills on sea surfaces, divided into three groups (Fig. 1): chemical, biological, and mechanical [2]. To address oil spills on sea surfaces, these procedures can be used individually or in combination. In situ burning, dispersant dispersion, and solidifier solidification are the chemical techniques [3]. While in situ burning can quickly eliminate oil spills, it can also produce harmful pollutants such as carbon dioxide, aromatic hydrocarbons, and nitrogen oxides, which can lead to secondary pollution in the atmosphere. Dispersive and solidification procedures remove oil contamination by reacting with it using surfactants or solidifiers, often using hazardous and unsustainable chemicals. The biological technique uses microorganisms like bacteria, algae, and fungi to naturally breakdown oil into non-toxic compounds [4]. It may cause secondary contamination in the marine environment by introducing exotic species. There are various ways for removing oils from water, such as oil booms, skimmers, chemical dispersants, membranes, bioremediation, mechanical recovery, and adsorption [5, 6]. Adsorption is an interesting technique for large-scale treatments due to its simplicity and low cost [7]. Using oil sorbents to handle oil spills in water is the most cost-effective and environmentally beneficial option since it allows for total collection and recovery of oil from water surfaces. The use of sorbents for oil spill cleanup is gaining interest, particularly for developing effective and eco-friendly options.

New materials for fast and effective oil removal from water are in demand due to possible impacts on ecosystems and human health. Sorbent materials are valuable for reducing oil spills as they can be used as a final step to clean up leftover oil in water after skimming [6]. Current oil sorbents are made from carbon [8–10], copolymers [11, 12], organoclays [13] and low-cost biodegradable materials [14–16], including porous structures [9, 12], films [10], fibers [15, 17, 18], and powders [14, 15]. Small sorbents are difficult to recover from water, limiting their reusability and increasing the expense of the operation. A novel method uses magnetic nanosorbents, which can be extracted from aquatic environments using magnetic separation [19]. Our interest in nanotechnology led us to develop nanosorbents with surface-modified magnetic particles for water pollution removal, including metal ions [20–24], organic dyes [25, 26], and pharmaceutics [27]. Limited research exists on magnetic nanosorbents for oil removal from water [28, 29].

Surface modification of magnetic particles is essential for achieving high adsorption capacities and achieving high affinity between magnetic sorbents and pollutant molecules. Here, magnetic sorbents with iron oxide nanoparticles in siliceous shells containing chitosan are mentioned for the first time. Chitosan is 2-glucosamine and N-acetyl-2-glucosamine units are obtained from deacetylation of chitin, the second-most common biopolymer seafood shells contain nature [27, 30]. Due to its characteristics, environmentally friendly, biodegradable, cheap, and chemically useful chitosan has garnered attention for wastewater treatment, particularly oil–water separation [31–34].

Since most oils float on the water surface, seabirds and marine animals are mainly affected [35] and other creatures and terrestrial animals are harmed if the oil comes ashore [36]. Inhalation, skin, and eye discomfort are also effects of such spilled oils on humans. Numerous researchers have analysed the aspects and effects of oil spills

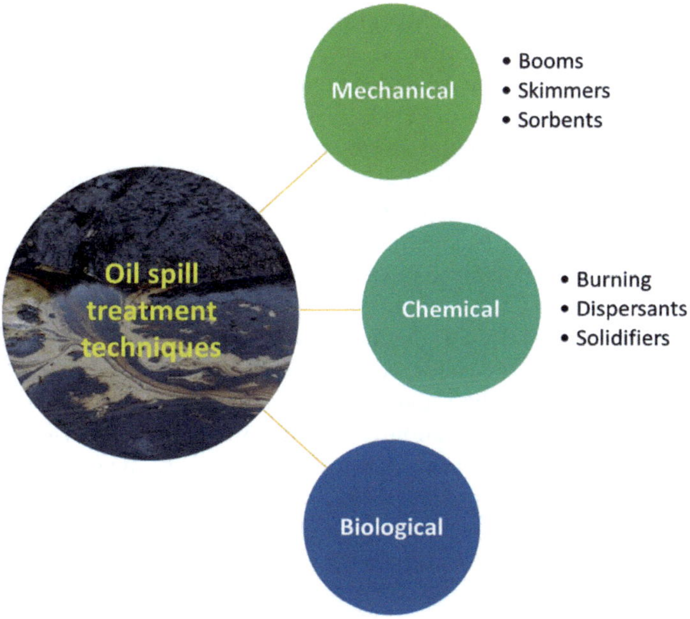


**Fig. 1** Treating oil spill techniques [42]

on the ecosystem over time [37–39] and devised numerous methods and chemicals for oil spill response [40, 41]. However, the intensity and location of the incident determine the cleanup technology that must be used.

Due to the diversity of the chitosan component, it has been predicted that a chitosan–silica composite would function more efficiently as an adsorbent than a pure silica material. It is widely recognized that chitosan possesses extraordinary sorption capabilities as a consequence of the polycationic structure of the molecule towards contaminants with anionic charge. On the other hand, it is a material that possesses low mechanical, thermal and chemical inertness or stability. In light of this, the combination of chitosan and silica, the function of which is to serve as a stiff support for the polymer due to its high stability, attempts to generate a composite that has the following properties: sorption properties as well as mechanical and thermal stability at a suitable level.

## 2 Synthesis of Chitosan-Silica Hybrid Nanoparticles

A silane precursor containing chitosan (Si Chitosan) was first synthesized for the purpose of the synthesis of the non-magnetic hybrid particles. This was accomplished by following a method that was previously published for the manufacture of a separate polysaccharide [43].

In order to produce magnetic hybrid nanoparticles ($Fe_3O_4$@$SiO_2$/Si Chitosan), the synthesis process has been divided over two independent phases [27]. The first step in the process involves the formation of the magnetic core (magnetite, or $Fe_3O_4$) through the alkaline hydrolysis of $FeSO_4 7H_2O$ in the presence of a stream of nitrogen [44]. After that, particles of $Fe_3O_4$ were encased in amorphous silica shells using the process of alkaline hydrolysis and condensation of a combination of tetraethyl orthosilicate (TEOS) and the precursor Si Chitosan. For the purpose of the encapsulation, a suspension of $Fe_3O_4$ nanoparticles (40 mg) in 38 mL of ethanol was prepared. This mixture was then maintained in an ice bath while being sonicated (with a horn, Sonics, or Vibracell).

After waiting 15 min, 2.4 mL of ammonia and a mixture of 0.406 mL of TEOS and 0.4 g of Si Chitosan precursor were slowly added to the solution. The mixture was then left for 2 h while being sonicated. In order to avoid the warming that would have been generated by the sonication, the reaction was carried out in an ice bath. The temperature of the reaction mixture, on average, was thirty degrees Celsius. After that, the particles that were produced were collected magnetically using a NdFeB magnet, and then they were completely cleaned with ethanol. At last, the particles were allowed to dry out due to the evaporation of the solvent.

## 3 Alteration of Nanocomposites Properties for Oil Spill Treatment

Adjusting the surface wettability of nanocomposites is crucial for their ability to absorb oil and resist water. Other characteristics of the material include increased oil uptake, retention, and capability to absorb oil, unsinkability, and reusability.

### *3.1 Oleophilicity and Hydrophobicity*

Oleophilicity and hydrophobicity refer to a material's ability to reject water and absorb oil. The selectivity of nanocomposite for oil and water interaction is linked to its wettability, which refers to the liquid's tendency to spread over solid surfaces. The shape of liquid drops on solid surfaces is determined by wettability. The wettability of a material is determined by the contact angle between its liquid interface and the solid surface. At contact angles below and above 90°, the surface becomes more hydrophilic or hydrophobic. When the contact angle of water or oil is less than or greater than 90 °, the surfaces of nanocomposites exhibit hydrophilic/oleophilic and hydrophobic/oleophilic characteristics [45, 46]. By altering the chemical composition and surface roughness of nanocomposites, the water contact angle can be increased from 0° to 150°, resulting in materials with improved hydrophobic and oleophilic properties. This means the nanocomposite's surface tension is between oil and water.

Nanocomposites can be altered to increase surface roughness in order to get super-hydrophobicity. Various organic polymers, including hydrophobic organosilanes, surfactants, and long-chain alkanes (thiol/fluorine/amine), have been employed to modify the surface of nanocomposites. In order to improve particular sorption behaviours, nanocomposites can be tailored by changing metal sites, functional groups, or surface changes. The sorption capabilities can be considerably enhanced by the addition of functional groups such as amines, –COOH, and –OH. To enhance stability, the structure might be changed either before or after synthesis.

## 3.2 *Porosity*

The empty gaps and pore structure significantly impact oil sorption. Measuring material uptake for effective selectivity towards a certain oil is crucial due to varying chemical compositions and pore shapes. Oil absorptivity is measured by measuring the weight growth of material in contact with oil over time. The porosity of the material significantly impacts oil absorption. Nanocomposites with high porosity are ideal for oil spill cleanup due to their increased oil absorption. Capillary force must be high enough to fill pores and connect spaces, ensuring oil stays in the pores.

The capillary force of a nanocomposite changes with pore size, increasing as pore size decreases and vice versa. Studies indicate that oil retention is impossible and maximal adsorption occurs at pores over 2 mm. The oil removal tool must be unsinkable to properly contact the oil and accomplish maximal removal. In addition, low-density materials are easier to collect after adsorption. Introducing magnetic-based biopolymeric or polymeric functionalized nanocomposites to a water–oil mixture causes them to float on the surface and become easily collected. The adsorption and recovery of these materials are shown in Fig. 2 [47]. The nanocomposite, with its hydrophobic character and low density, repels water after being introduced to the surface-foasting oil [48, 49].

## 3.3 *Recovery and Reusability*

Material recovery and reusability are the main concerns for their applicability. If adsorbents cannot be recovered after adsorption, the cost may grow, potentially leading to secondary pollution. Enhanced reusability and sorption capacity can be achieved by studying and implementing new materials in synthesis methods. Research suggests that putting nanocomposites on substrates like sponges and foams can improve separation performance. The use of forams and sponges as nanocomposite substrates has gained popularity due to their ability to adjust pore size and

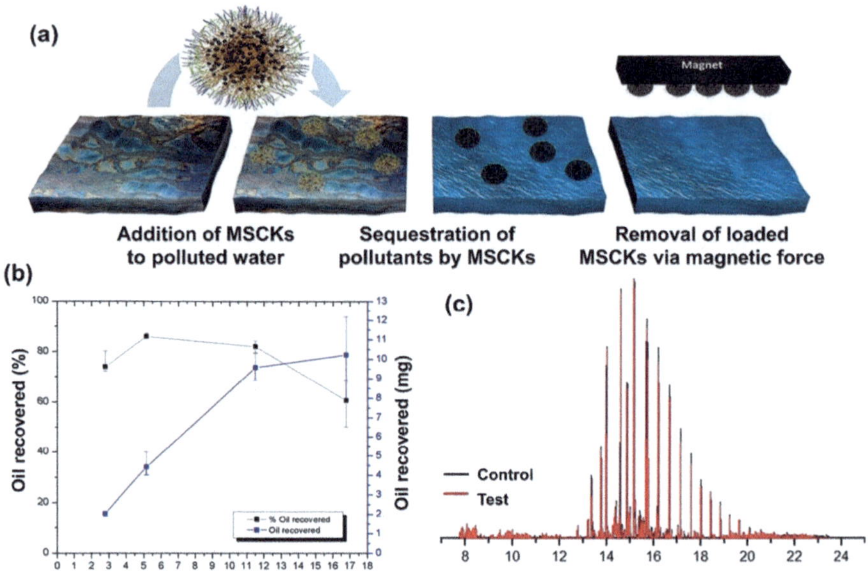


**Fig. 2** **a** The process of oil sequestration from the polluted stream using floatable modified magnetic nanocomposites (magnetic shell cross-linked kneel-like nanoparticles (MSCKs). **b** Oil recovery data. **c** Gas chromatography–mass spectrometry of oil extracted from control (black) and test (red) groups, Reproduced from [47]

aesthetics. Sorbed oil can be recovered using vacuum filtration, distillation, or extraction.

To reuse sorbent materials, chemical and mechanical stability, as well as oleophilic properties, are crucial. High-pressure oil recovery can distort materials, reducing their sorption capacity if they are not mechanically stable. Distillation and extraction methods require thermal and chemical stability [50, 51].

Yu et al. and Qiao et al. [46, 49] found that recovery is simple for materials with ferromagnetic characteristics. Recovering magnetic nanocomposites from oil-polluted areas is simple using an external magnetic field [52–54]. Despite decreased saturation magnetization compared to bare magnetic NPs, nanocomposite can be easily recovered in seconds using an external magnetic field, simplifying and reducing recovery costs [48, 49].

Another benefit of nanocomposites is their reusability after oil adsorption. The primary regeneration method is mechanical squeezing or ultrasonic washing in hexane or ethanol [55]. One benefit of reusing oil sorbents is reduced cleanup costs and recovery. Regenerated materials can be reused without affecting oil adsorption capacity or contact angle. Studies suggest nanocomposites can be reused up to 1,000 times [45, 56].

## 4 Oil Adsorption

Chitosan is the biopolymer that is used most frequently when applied to the synthesis of nanocomposites. Chitosan (CS) has been utilized in the process of modifying the MNPs in order to synthesize oil sorbents. As silane coupling agents, aminopropyltriethoxysilane (APTES) [57, 58] and 3-(triethoxysilyl)propyl isocyanate (ICPTES) [59] were used to create MNPs coated with silica and grafted with chitosan. According to Soares et al. [59], the grafting of chitosan onto an inorganic matrix takes place via urea bonds. This is because the $NH_2$ groups of chitosan react more quickly than the OH groups. Because of their electrostatic attraction, CS-grafted MNPs have the potential to focculate oil droplets in environments that are acidic or neutral.

Alkaline circumstances overcome electrostatic repulsion and hydrophobic contact, causing MNPs to adsorb over oil droplets, suggesting a demulsification effect under magnetic fields [57, 58]. APTES-coated MNPs only improved oil separation efficiency in the acidic range, but quaternized MNPs performed well at varied pH levels.

Chitosan reinforced with GO-polydopamine aerogel, modified with perfuorodecanethiol synthesized from self-polymerizing dopamine on GO sheets, showed super-hydrophobic properties (Fig. 3). The nanocomposite aerogel exhibits superoleophobicity in water and super-amphiphilicity in air without modification [60]. HPDCS, a branched-modified chitosan polymer, showed superior thickening and shear resistance, and had a 24.12% oil recovery potential at 2000 mg/L concentration compared to HPAM [61].

As sample adsorbates, toluene, cyclohexane, n-heptane, and chloroform that had been dyed with Nile red were chosen. In this case, the oil-like organic solvent (0.1 mL) was put on top of the water in a petri dish. Then, about 3.4 mg of nanoparticles made of $Fe_3O_4@SiO_2$/Si Chitosan were carefully put on the surface of the oil–water mixtures. After the oil had completely covered the nanoparticles, the $Fe_3O_4@SiO_2$/Si Chitosan particles and the oil that had stuck to them were collected with a NdFeB magnet and weighed.

## 5 Desorption and Reuse

Nanoparticles were tested for oil desorption and reusability in toluene. In order to desorb toluene, magnetic nanoparticles were collected, cleaned with 10 mL of ethanol using ultrasounds, and then magnetically separated. The process was continued until no color was visible in the supernatant, usually four times. At room temperature, particles dried. Reusing nanoparticles after desorption was employed in adsorption tests to test their reusability. Weight gain was calculated after four repetitions of the technique.

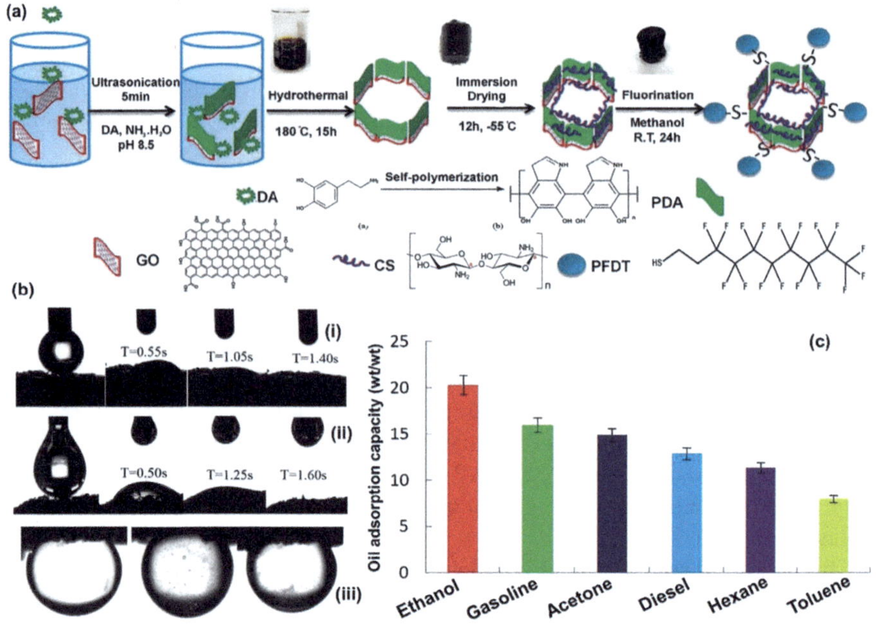


**Fig. 3** **a** Illustrates the process of forming chitosan/reduced graphene oxide composite aerogel (CGA) and perfluorinated superhydrophobic polydopamine/chitosan/reduced graphene oxide composite aerogel (fCGA); **b** the spreading process of a droplet of water (i) and diesel (ii) on the CGA surface; optical images of a diesel droplet on the CGA surface in NaOH, HCl and NaCl aqueous solutions (iii); **c** contact angle of various organic liquids under the water of CGA. Reproduced from [59]

# 6 Application of Chitosan-Silica Hybrid Nanomaterials for Oil Spill Treatment

Oil pollution and spill cleanup costs billions and involves hundreds or thousands of workers. Low-, medium-, and high-density oil wastes are hazardous and can harm aquatic plants, animals, and humans. The petrochemical, metallurgical, pharmaceutical, and food industries discharge massive amounts of oil waste annually, posing serious environmental risks [62]. Oil waste is divided into crude oil and polyaromatic hydrocarbons (PAHs) like diesel, petrol and petroleum ether. The type and amount of oil exposed determine oil waste's environmental toxicity. Animal fur and feathers lose their insulation to oil waste, causing hypothermia and swimming impairment that can kill oiled animals. Touching, inhaling, and eating polluted food from oil wastes harms humans and aquatic life. Several papers describe oil waste risks. Figure 4 shows the literature review report on the notable toxicity of oil waste and oil spill.

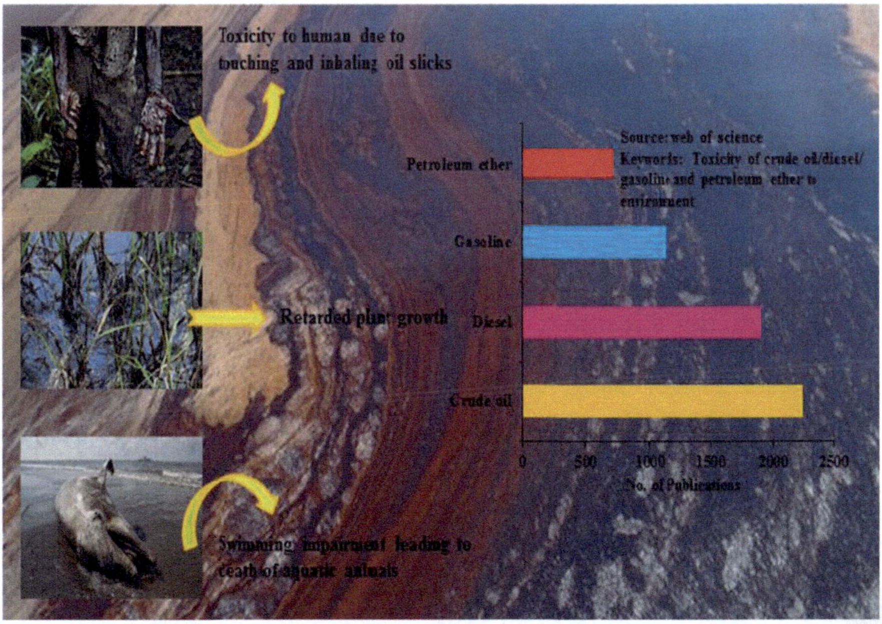


**Fig. 4** "Toxicity of oil wastes/oil spills to the environment with their number of published reports in response to crude oil, diesel, and gasoline and petroleum ether wastes." Reproduced from [63]

Due to their small pores, their preference for fat, and their aversion to water, chitosan–silica hybrid aerogels have the potential to be utilised for oil absorption. By measuring the mass of the hybrid aerogels after they had absorbed oil at predetermined intervals, we were able to determine how well the hybrid aerogels performed oil absorption. This demonstrated the process by which the oils were absorbed and consumed. Chemical solvents such as petrol, arachis oil, pentane, hexane, acetone, dichloromethane, xylene, and hexane were among the substances that were utilised in this process. For the chitosan–silica composite aerogel with 20% weight percent chitosan (20-CS), the oil-absorbing curves are shown in Fig. 5. The first two minutes saw a fast absorption of organic liquids. After that, the rate of absorption slowed down, and within 10 min, the oil absorbency reached its peak. Because aerogels are made up of tiny pores, they have a capillary force that pulls in liquids.

**Fig. 5** **a** "Photographs of the composite aerogels in the water phase and oil/water mixture and **b** oil absorption dynamic curves of 20-CS." Reproduced from [64]

(a)

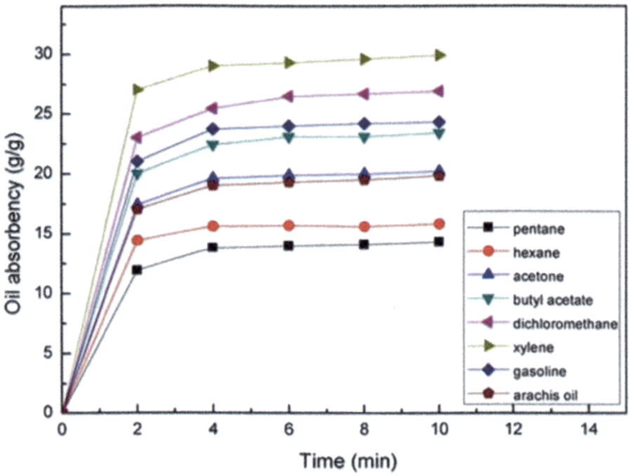


(b)

## 7 Conclusion

In general, hybrid nanomaterials consisting of chitosan and silica show great promise in treating oil spills due to their adaptability, high adsorption capacity, and selectivity. Their synthesis methods, characteristics, and uses show how they can be used to help with environmental sustainability and solve the problems caused by oil spills. With increased research and development, these nanomaterials could become essential components of future oil spill response approaches, preventing marine ecosystems and coastal communities around the world.

## References


1. Ge J, Zhao HY, Zhu HW, Huang J, Shi LA, Yu SH (2016) Advanced sorbents for oil-spill cleanup: Recent advances and future perspectives. Adv Mater 28:10459–10490
2. Prendergast DP, Gschwend PM (2014) Assessing the performance and cost of oil spill remediation technologies. J Clean Prod 78:233–242
3. Adofo YK, Nyankson E (2022) Dispersants as an oil spill clean-up technique in the marine environment: a review. Heliyon 8:e10153
4. Azubuike CC, Chikere CB, Okpokwasili GC (2016) Bioremediation techniques-classification based on site of application: principles, advantages, limitations and prospects. World J Microbiol Biotechnol 32:180
5. Ventikos NP, Vergetis E, Psaraftis HN, Triantafyllou G (2004) A high-level synthesis of oil spill response equipment and countermeasures. J Hazard Mater 107:51–58
6. Dave D, Ghaly AE (2011) Remediation technologies for marine oil spills: a critical review and comparative analysis. Am J Environ Sci 7:423–440
7. Taka AL, Pillay K, Mbianda XY (2017) Nanosponge cyclodextrin polyurethanes and their modification with nanomaterials for the removal of pollutants from waste water: a review. Carbohydr Polym 159:94–107
8. Fard AK, Rhadfi T, Mckay G, Al-marri M, Abdala A, Hilal N, Hussien MA (2016) Enhancing oil removal from water using ferric oxide nanoparticles doped carbon nanotubes adsorbents. Chem Eng J 293:90–101
9. Bi H, Xie X, Yin K, Zhou Y, Wan S, Ruoff RS, Sun L (2014) Highly enhanced performance of spongy graphene as an oil sorbent. J Mater Chem A 2:1652–1656
10. Feng C, Yi Z, She F, Gao W, Peng Z, Garvey CJ, Dumee LF, Kong L (2016) Superhydrophobic and superoleophilic micro-wrinkled reduced graphene oxide as a highly portable and recyclable oil sorbent. ACS Appl Mater Interfaces 8:9977–9985
11. Zhang T, Guo Q (2017) Continuous preparation of polyHIPE monoliths from ionomerstabilized high internal phase emulsions (HIPEs) for efficient recovery of spilled oils. Chem Eng J 307:812–819
12. Lin P, Meng L, Huang Y, Liu L (2016) Synthesis of porous polyurea monoliths assisted by centrifugation as adsorbents for water purification. Colloids Surfaces A Physicochem Eng Asp 506:87–95
13. Adebajo MO, Frost RL, Kloprogge JT, Carmody O (2003) Porous materials for oil spill cleanup: A Review of synthesis and absorbing properties. J Porous Mater 10:159–170
14. Zang D, Liu F, Zhang M, Gao Z, Wang C (2015) Novel superhydrophobic and superoleophilic sawdust as a selective oil sorbent for oil spill cleanup. Chem Eng Res Des 102:34–41
15. Zhang Y, Yang S, Wu JQ, Yuan TQ, Sun RC (2014) Preparation and characterization of lignocellulosic oil sorbent by hydrothermal treatment of populus fiber. Materials (Basel). 7:6733–6747
16. Bazargan A, Tan J, Hui CW, McKay G (2014) Utilization of rice husks for the production of oil sorbent materials. Cellulose 21:1679–1688
17. Hussein M, Amer AA (2011) Sawsan II heavy oil spill cleanup using low grade raw cotton fibres: trial for practical application. J Pet Technol Altern Fuels 2:132–140
18. Choi HM, Moreau JP (1993) Oil sorption behaviour of various sorbents studied by sorption capacity measurement and environmental scanning electron microscopy. Microsc Res Tech 25:447–455
19. Mehta D, Mazumdar S, Singh SK (2015) Magnetic adsorbents for the treatment of water/wastewater—a review. J Water Process Eng 7:244–265
20. Carvalho RS, Daniel-da-Silva AL, Trindade T (2016) Uptake of Europium(III) from water using magnetite nanoparticles. Part Part Syst Charact 33:150–157
21. Girginova PI, Daniel-da-Silva AL, Lopes CB, Figueira P, Otero M, Amaral VS, Pereira E, Trindade T (2010) Silica coated magnetite particles for magnetic removal of $Hg^{2+}$ from water. J Colloid Interface Sci 345:234–240

22. Lopes CB, Figueira P, Tavares DS, Lin Z, Daniel-da-Silva AL, Duarte AC, Rocha J, Trindade T, Pereira E (2013) Core–shell magnetite-silica dithiocarbamate-derivatised particles achieve the water framework directive quality criteria for mercury in surface waters. Environ Sci Pollut Res Int 20:5963–5974
23. Pinheiro PC, Tavares DS, Daniel-da-Silva AL, Lopes CB, Pereira E, Araújo JP, Sousa CT, Trindade T (2014) Ferromagnetic sorbents based on nickel nanowires for efficient uptake of mercury from water. ACS Appl Mater Interfaces 6:8274–8280
24. Tavares DS, Daniel-da-Silva AL, Lopes CB, Silva NJO, Amaral VS, Rocha J, Pereira E, Trindade T (2013) Efficient sorbents based on magnetite coated with siliceous hybrid shells for removal of mercury ions. J Mater Chem A 1:8134
25. Salgueiro AM, Daniel-da-Silva AL, Girão AV, Pinheiro PC, Trindade T (2013) Unusual dye adsorption behavior of κ-carrageenan coated superparamagnetic nanoparticles. Chem Eng J 229:276–284
26. Daniel-da-Silva AL, Salgueiro AM, Creaney B, Oliveira-Silva R, Silva NJO, Trindade T (2015) Carrageenan-grafted magnetite nanoparticles as recyclable sorbents for dye removal. J Nanoparticle Res 17:302
27. Soares SF, Simões TR, António M, Trindade T, Daniel-da-Silva AL (2016) Hybrid nanoadsorbents for the magnetically assisted removal of metoprolol from water. Chem Eng J 302:560–569
28. Chen M, Jiang W, Wang F, Shen P, Ma P, Gu J, Mao J, Li F (2013) Synthesis of highly hydrophobic floating magnetic polymer nanocomposites for the removal of oils from water surface. Appl Surf Sci 286:249–256
29. Tao S, Ding Y, Jiang L, Li G (2016) Preparation of magnetic hierarchically porous microspheres with temperature-controlled wettability for removal of oils. J Colloid Interface Sci 492:73–80
30. Rafique A, Zia KM, Zuber M, Tabasum S, Rehman S (2016) Chitosan functionalized poly(vinyl alcohol) for prospects biomedical and industrial applications: a review. Int J Biol Macromol 87:141–154
31. Sokker HH, El-Sawy NM, Hassan MA, El-Anadouli BE (2011) Adsorption of crude oil from aqueous solution by hydrogel of chitosan based polyacrylamide prepared by radiation induced graft polymerization. J Hazard Mater 190:359–365
32. Ahmad AL, Sumathi S, Hameed BH (2005) Adsorption of residue oil from palm oil mill effluent using powder and flake chitosan: equilibrium and kinetic studies. Water Res 39:2483–2494
33. Zhang S, Lü T, Qi D, Cao Z, Zhang D, Zhao H (2017) Synthesis of quaternized chitosan-coated magnetic nanoparticles for oil-water separation. Mater Lett 191:128–131
34. Liu J, Li P, Chen L, Feng Y, He W, Yan X, Lu X (2016) Superhydrophilic and underwater superoleophobic modified chitosan-coated mesh for oil/water separation. Surf Coat Technol 307:171–176
35. Barron MG (2012) Ecological impacts of the deepwater horizon oil spill: implications for immunotoxicity. Toxicol Pathol 40:315–320
36. Brody TM, Bianca DP, Krysa J (2012) Analysis of inland crude oil spill threats, vulnerabilities, and emergency response in the Midwest United States. Risk Anal 32:1741–1749
37. Jernelöv A (2010) The Threats from oil spills: now, then, and in the future. Ambio 39:353–366
38. Henkel JR, Sigel BJ, Taylor CM (2012) Large-scale impacts of the deepwater horizon oil spill: can local disturbance affect distant ecosystems through migratory shorebirds? Biosci 62:676–685
39. Chang SE, Stone J, Demes KW, Piscitelli-Doshkov MA (2014) Consequences of oil spills: a review and framework for informing planning. Ecol Soc 19:26
40. TOPF (2014) ITOPF Response Techniques [Online]
41. Michel J, Shigenaka G, Hoff RZ (1992) Oil spill response and cleanup techniques. An Introduction to coastal habitats and biological resources for oil spill response. Publisher, NOAA/Hazardous Materials Response and Assessment Division, pp 1–103
42. Ouyang D, Lei X, Zheng H (2023) Recent advances in biomass-based materials for oil spill cleanup. Nanomaterials 13:620

43. Soares SF, Trindade T Daniel-da-Silva AL (2015) Carrageenan-silica hybrid nanoparticles prepared by a non-emulsion method. Eur J Inorg Chem 4588–4594
44. Oliveira-Silva R, Pinto da Costa J, Vitorino R, Daniel-da-Silva AL (2015) Magnetic chelating nanoprobes for enrichment and selective recovery of metalloproteases from human saliva. J Mater Chem B 3:238–249
45. Pinto J, Athanassiou A, Fragouli D (2018) Surface modification of polymeric foams for oil spills remediation. J Environ Manage 206:872–889
46. Qiao K, Tian W, Bai J, Wang L, Zhao J, Du Z, Gong X (2019) Application of magnetic adsorbents based on iron oxide nanoparticles for oil spill remediation: a review. J Taiwan Inst Chem Eng 97:227–236
47. Pavia-Sanders A, Zhang S, Flores JA, Sanders JE, Raymond JE, Wooley KL (2013) Robust magnetic/polymer hybrid nanoparticles designed for crude oil entrapment and recovery in aqueous environments. ACS Nano 7:7552–7561; Iftekhar S, Deb A, Heidari G, Sillanpää M, Lehto VP, Doshi B, Hosseinzadeh M, Zare EN (2023) A review on the effectiveness of nanocomposites for the treatment and recovery of oil spill. Environ Sci Pollut Res 30:16947–16983
48. Zhu Q, Tao F, Pan Q (2010) Fast and selective removal of oils from water surface via highly hydrophobic core– shell $Fe_2O_3$@ C nanoparticles under magnetic feld. ACS Appl Mater Interfaces 2:3141–3146
49. Yu L, Hao G, Gu J, Zhou S, Zhang N, Jiang W (2015) Fe3O4/PS magnetic nanoparticles: synthesis, characterization and their application as sorbents of oil from waste water. J Magn Magn Mater 394:14–21
50. Laitinen O, Suopajärvi T, Sirviö JA, Liimatainen H (2019) Superabsorbent aerogels from cellulose nanofibril hydrogels In: Mondal M (eds) Cellulose-based superabsorbent hydrogels. Polymers and polymeric composites: a reference series. Springer Cham. https://doi.org/10.1007/978-3-319-77830-3_20
51. Shahmirzaee M, Abdi J, Hemmati-Sarapardeh A, Schaffie M, Ranjbar M, Khataee A (2022) Metal-organic frameworks as advanced sorbents for oil/water separation. J Mol Liq 363:119900
52. Akter T, Bañuelos JL, Andrade D, Banuelos DI, Saupe GB (2022) Rapid adsorption mechanism of methylene blue onto a porous mixed Ti-Nb oxide. Mater Chem Horizons 1:49–67
53. Hassanzadeh-Afruzi F, Heidari G, Maleki A (2022) Magnetic nanocomposite hydrogel based on arabic gum for remediation of lead(II) from contaminated water. Mater Chem Horizons 1:107–122
54. Mehdizadeh A, Moghadam PN, Ehsanimehr SS, Fareghi AR (2022) Preparation of a new magnetic nanocomposite for the removal of dye pollutions from aqueous solutions: synthesis and characterization. Mater Chem Horizons 1:23–34
55. Gui X, Zeng Z, Lin Z, Gan Q, Xiang R, Zhu Y, Cao A, Tang Z (2013) Magnetic and highly recyclable macroporous carbon nanotubes for spilled oil sorption and separation. ACS Appl Mater Interfaces 5:5845–5850
56. Liu J, Wang H, Li X, Jia W, Zhao Y, Ren S (2017) Recyclable magnetic graphene oxide for rapid and efcient demulsifcation of crude oil-in-water emulsion. Fuel 189:79–87
57. Zhang S, Lü T, Qi D, Cao Z, Zhang D, Zhao (2017) Synthesis of quaternized chitosancoated magnetic nanoparticles for oil-water separation. Mater Lett 191:128–131
58. Soares SF, Rodrigues MI, Trindade T, Daniel-da-Silva AL (2017) Chitosan-silica hybrid nanosorbents for oil removal from water. Colloids Surf A Physicochem Eng Asp 532:305–313
59. Cao N, Lyu Q, Li J, Wang Y, Yang B, Szunrits S, Boukherroub R (2017) Facile synthesis of fluorinated polydopamine/chitosan/reduced graphene oxide composite aerogel for efficient oil/water separation. Chem Eng J 326:17–28; Iftekhar S, Deb A, Heidari G, Sillanpää M, Lehto VP, Doshi B, Hosseinzadeh M, Zare EN (2023) A review on the effectiveness of nanocomposites for the treatment and recovery of oil spill. Environ Sci Pollut Res 30:16947–16983
60. Liu Y, Lai N (2019) Synthesis and properties of branched polymer based on modifed chitosan for enhanced oil recovery. IOP Conf Ser: Earth Environ Sci 267:022038
61. Chen W, Su Y, Zheng L, Wang L, Jiang Z (2009) The improved oil/water separation performance of cellulose acetate-graft-polyacrylonitrile membranes. J Membr Sci 337(1–2):98–105

62. Salman M, Jahan S, Kanwal S, Mansoor F (2019) Recent advances in the application of silica nanostructures for highly improved water treatment: a review. Environ Sci Pollut Res 26:21065–21084
63. Ma Q, Liu Y, Dong Z, Wang J, Hou X (2015) Hydrophobic and nanoporous chitosan–silica composite aerogels for oil absorption. J Appl Polym Sci 132(15) 41770
64. Nguyen ST, Feng J, Le NT, Le ATT, Hoang N, Tan VBC, Duong HM (2013) Cellulose aerogel from paper waste for crude oil spill cleaning. Ind Eng Chem Res 52:18386–18391

# Microbial Biosurfactants for Oil Spill Treatment

**Hauwa A. Umaru, Micheal Sunday Abu, Kerenhappuch Isaac Umaru, and Isaac John Umaru**

**Abstract** The remediation agent used for oil spill clean-up such as chelating and sodium lauryl sulphate or acetytrimethyl ammonium bromide as complexion has become the common environmental oil clean-up. However, efficiently this agent is observed to be non- biodegradable and may cause secondary contaminants. The surfactants are reported as a group of amphophilic chemical compound of both hydrophobic and hydrophilic characteristics which form indispensable components in the industry. The importance is evidence from the enormous amount used and its consistent utilization, ranging from food and beverage, in agriculture, public health as well as bioremediation and textile industries.

## 1 Introduction

Over the last decade, biosurfactants for oil and gas applications have garnered significant attention and study, therefore this chapter provides an update and state-of-the-art on this burgeoning topic [1]. First, there are two main reasons for this interest rise. One is that the oil production rate, which also accounts for the pace of finding new subterranean oil sources, has been declining for decades. Global subterranean oil reserves are limited; hence production must reach a peak (Hubbard peak), which it has [2]. This rate has now decreased, which has generated interest in employing biosurfactants to recover residual crude oil using microbial enhanced oil recovery using microorganisms or their products. The dispute over synthetic chemical surfactants in chemical dispersion formulations is another major cause for this spike in interest in producing biosurfactants for the Oil and Gas sector.

Chemical remediation utilizing chelating agents and surfactants such sodium lauryl sulfate or acetyl trimethyl ammonium bromide as complexing has replaced

H. A. Umaru · M. S. Abu · K. I. Umaru · I. J. Umaru (✉)
Department of Biochemistry, Moddibo Adama University, Yola, Nigeria
e-mail: umaruisaac@gmail.com

Department of Biochemistry, Federal University Wukari, Taraba, Nigeria

© The Author(s), under exclusive license to Springer Nature Singapore Pte Ltd. 2025

S. A. Bhawani et al. (eds.), *Biobased Materials and Their Composites for Oil Spill Treatment*, Composites Science and Technology,
https://doi.org/10.1007/978-981-96-6049-0_10

environmental oil spill cleanup [3]. Although efficient, these chemicals are non-biodegradable and may cause secondary contaminants [4]. Therefore, natural surfactant alternatives and eco-friendly usage are needed. Thus, microbial surface-active chemicals (MSAC) may be alternatives. These microbes create chemicals that solubilize hydrophobic substrates such hydrocarbons, lipids, oils, and antibiotics, making them carbon sources [5]. MSAC's foam, emulsifier, and solubilizing properties enable it to solubilize, disperse, and desorb priority environmental contaminants such petroleum derivatives, PAHs, and heavy metals [6].

Surfactants are a group of amphiphilic chemical compounds (i.e., having both hydrophobic and hydrophilic domains) that form an indispensable component in almost every sector of modern industry. Their significance is evidenced from the enormous volumes that are used and wide diversity of applications they are used in, ranging from food and beverage, agriculture, public health, healthcare/medicine, textiles, and bioremediation [1]. Amphiphilic chemicals such as biosurfactants have the ability to lower interfacial tension, disperse oil particles, and break them down into non-toxic debris. Rhamnolipid, sophorolipid, and surfactin are three of the many biosurfactants that have been thoroughly studied and only Pseudomonas, Bacillus, and Candida have been studied among the microorganisms; other potentially useful candidates may exist [7]. One way to measure a surfactant's efficacy is by looking at how well it reduces the surface tension (ST) and interfacial tension (IFT) between two immiscible phases. The surface tension (ST) is a density-based metric describing the amount of energy needed to raise a liquid's surface area as a result of interactions between its molecules. A lower surface tension (ST) is achieved when a surfactant is present because fewer efforts are needed to get a molecule to the surface. According to Mulligan [8], a reliable biosurfactant may reduce the surface tension of water from 72 to less than 35 mN/m and the interfacial tension of water against n-hexadecane from 40 to less than 1 mN/m.

One kind of MSAC that finds widespread use in green cleaning products is biosurfactants. Different kinds of bacteria, yeast, fungus, and archaea may manufacture these secondary metabolites from renewable energy sources; they have antimicrobial, antifungal, and antiviral characteristics [9]. They work well in a variety of environments with varying temperatures, pH, and salinity, and they are biodegradable and have minimal toxicity [3, 10]. Because they improve the bioavailability and solubility of hydrophobic contaminants, biosurfactants have the makings of promising bioremediation candidates [11]. Biosurfactants have shown to be effective complexing agents for heavy metals and other contaminants by creating micelles bound to the metal ions inside. Furthermore, these contaminants are more easily mobilized and recovered due to biosurfactant binding activity and their broad metal selectivity spectrum [12].

Classification of Surfactants

The hydrophilic heads of surfactants are categorized as anionic, cationic, amphoteric, or uncharged based on their charge, which affects their solubility. Some examples of anionic surfactants include sodium dodecyl sulfate (SDS) and linear alkylbenzene sulfonate (LAS), both of which are common ingredients in laundry detergents and other personal and home cleaning products [13]. Additionally, anionic surfactants are used extensively in the cosmetic and pharmaceutical industries, as well as in the removal of hydrocarbons from soil. Detergents, fabric softeners, and hair conditioners all make use of cationic surfactants. And because of their bacteriostatic properties, they may be used to create antimicrobial and cleaning agents [13], Depending on the surrounding conditions and pH, amphoteric surfactants like lauryl betaine may exhibit anionic or cationic characteristics. As emulsifiers, wetting agents, and foam stabilizers, uncharged surfactants find extensive use.

Microbial Surface-Active Compounds

Unique molecular compositions may be produced by microorganisms in the form of biosurfactants. A biosurfactant's molecular weight determines whether it belongs to the high or low molecular weight category. High molecular weight biosurfactants include lipoprotein, lipopolysaccharide, protein, polysaccharide, and biopolymer complexes. Low molecular weight biosurfactants have been described as those based on lipopeptides, glycolipids, and phospholipids [14]. Based on their chemical compositions, biosurfactants derived from microbes may be classified as glycolipids, lipopeptides and lipoproteins, phospholipids, or polymeric surfactants. Sophorolipids, rhamnolipids, and trehalolipids are the primary glycolipid-type biosurfactants. Most notably, Pseudomonas aeruginosa produces rhamnolipids. There have been reports of its antibacterial and antifungal activities (Benincasa et al., 2004), in addition to its ability to emulsify hydrocarbons. In medicine, trehalolipids are used for immunomodulatory and anticancer purposes. They are also employed in bioremediation, microbial improved oil recovery, cosmetics, and food as hydrocarbon solubilizers and emulsifiers [15].

Classes of Biosurfactants and Their Producers

| Biosurfactants | | Microorganisms |
|---|---|---|
| Group | Biosurfactant type | Pseudomonas |
| Glycolipids | Rhamnolipids | Aeruginosa |
| | Trehalolipids | Mycobacterium |
| | | Tuberculosis |
| | | Rhodococcus |
| | | erythropolis |
| | | Arthrobacter sp |
| | | Nocardia sp |
| | | Corynebacterium sp |
| | Sophorolipids | Torulopsis |
| | | Bombicola |
| | | Torulopsis |
| | | Petrophilum |
| | | Torulopsis apicola |

(continued)

(continued)

| Biosurfactants | | Microorganisms |
|---|---|---|
| Group | Biosurfactant type | Pseudomonas |
| Fatty acids, Phospholips and Neutral fats | Carinomycolic acid | Corynebacterium lepus |
| | Spiculisporic acid | Penicillium spiculisporum |
| | Phosphatidyl ethanolamine | Acinetobacter sp |
| | | Rhodococcus erythropolis |
| Lipopeptides | Surfactin | Bacillus subtilis |
| | Lichenysin | Bacillus licheniformis |
| Polymeric biosurfactants | Emulsan | Acinetobacter calcoaceticus |
| | Alasan | Acinetobacter radioresistens |
| | Liposan | Candida lipolytica |

Adopted from Gürkök & Özdal [16]

Biosurfactant-Producing Microorganisms

P. saureans

Gammaproteobacteria Pseudomonas produce biosurfactants such rhamnolipids and lipopeptides like viscosin, amphisin, tolaasin, and syringomycin. Pseudomonas has numerous marine species, although most are land-based [17]. Pseudomonas aeruginosa is the most studied biosurfactant producer. It grows well on hydrocarbon and non-hydrocarbon substrates and produces stable rhamnolipid emulsions with crude oil and kerosene [18].

Bacillus

Bacillus, which produces biosurfactants for mesocosms, is found near oil reservoirs or oil-polluted soils. After extraction from a petroleum reservoir, Bacillus methylotrophicus USTBa grew well in an aqueous medium with crude oil. Over 90% of crude oil was extracted by B. methylotrophicus after 12 days. According to Chandankere et al. [19], the bacteria produced a strong glycolipid biosurfactant with a ST of 28 mN/m in the growing medium.

E. coli

Acinetobacter bacteria are aerobic, gram-negative. Gammaproteobacteria from Pseudomonadales. Marine habitats and the rest of nature are home to Acinetobacter. This genus produces hydrocarbon-degrading extracellular EPS [20].

Anaerobic bacteria

The order Rhodobacterales includes strictly aerobic gram-negative Antarctobacter bacteria. Labrenz et al. [21] found Antarctobacter heliothermus to be the sole Antarctic-isolated species with a valid taxonomic description. The marine bacterium Antarctobacter sp. strain TG22 produces an AE22-type extracellular water-soluble glycoprotein polymer. This polymer formed stable emulsions with vegetable oils at 0.02% [22].

Coliforms

Rhodococcus species have diverse metabolisms and can thrive in many settings [23]. Wang et al. [24] studied the genus' ability to digest hydrocarbons and pollutants from varied habitats. Rhodococcus ruber, aurantiacus, and erythropolis produce biosurfactants [25]. N-alkanes were the sole way R. erythropolis 3C-9 could grow and produce biosurfactant (CMC of 50 mg/L), while glucose had no impact. The 3C-9 biosurfactant had two glycolipids, primarily glucose and trehalose monosaccharides, and C10–C22 fatty acids. Docosenoic and hexadecenoic acids predominated.

Hydrocarbon enrichment causes halomonas family members to react [26] and creates useful EPS [22]. WB1, a thermophilic H. nitroreducens strain from a hot spring, bound metals and emulsified vegetable oils (68–85%). It also dissolved 56–65% aliphatic hydrocarbons. WB1's EPS contained glucose, galactose, mannose, and uronic acid [27]. Halomonas eurihalina strain H96, isolated from saline soil in Spain, has a high uronic acid content in its extracellular polymer (EPS) [28], similar to several marine strains [22]. Besides emulsifying, certain halophilic Halomonas species produce anticancer highly sulphated exopolysaccharides [29]. T cells from acute lymphoblastic leukemia have a particular proapoptotic response to halophilic H. stenophila strain B100 [30].

Candida albicans

Pseudoalteromonas, a genus in the order Alteromonadales, produces glycolipid-type extracellular polysaccharides (EPS) with a wide range of biological activities and chemical composition and is often found in cold waters and sea ice [31]. The Arctic sea ice-isolated Pseudoalteromonas sp. strain SM20310 produced 567 mg/L EPS from mannose and glucose. The ecological role of the EPS improved the strain's high-salinity and low-temperature tolerance, according to Liu et al. [32].

Marinobacter

The Alteromonadales family includes Marinobacter. M. hydrocarbonoclasticus and M. algicola are usually isolated from oil-rich maritime environments [33]. Marinobacter can thrive and produce EPS on glucose and hydrocarbons, according to research.

Microbial biosurfactants in oil spill treatment

Bioremediation of petroleum soils and waters

Although sluggish, 60–90% of petroleum compounds biodegrade. Biosurfactants increase pollutant bioavailability, which aids oil cleanup [35]. Martins & Martins [36] found that hydrocarbon-contaminated bacteria may produce biosurfactants from petroleum sludge or garbage. Pseudomonas species create most biosurfactants in hydrocarbon-polluted settings, according to Naeem & Qazi [37]. In recent investigations, Prakash et al. [38] identified marine microbes such B. subtilis AS2, Bacillus licheniformis AS3, and Bacillus velezensis AS4 in this habitat. They can create biosurfactants with high emulsification values for low molecular weight hydrocarbons and crude oil degradation efficiency of 88, 92, and 97%. Since petroleum is a complex combination, several writers have utilized total petroleum hydrocarbon (TPH) reduction to compare biodegradation efficiency studies.

An approach to bioremediation of petroleum that utilizes biostimulation

Biostimulation uses a commercial or homemade biosurfactant (or fermentation extract) alone or in conjunction with additional substances [39]. Popular bioremediation technology biostimulation employs biosurfactants and biodegrades 50–97%. The main biosurfactants in biostimulation are rhamnolipids and lipopeptides. Rhamnolipids, biochar, and nitrogen fertilizer increased oil-degrading microorganisms and saline mash soil fertility [40]. Using rhamnolipid biostimulation at 1.5 g/L for six months, heterotrophic bacteria removed benz(d)anthracene, benz(b)fluorine, and fluoranthene with 80.5–86.1% TPH removal efficiency [41]. For oil bioremediation, most biostimulation experiments compared biosurfactants against commercial surfactants. At 40–80% CMC concentrations, C. sphaerica and Bacillus sp. biosurfactant crude extract (93 and 43%) cleaned motor oil from sand better than Tween 80 and Triton X-100 [42]. Ayed et al. [43] compared Tween 80 and SDS to Bacillus amyloliquefaciens An6 biosurfactants (0.05–2 g/L) for diesel oil removal. The biosurfactant emulsifies diesel and solubilizes it at 1 g/L with 71.54% efficiency.

Through pseudosolubilization, large quantities of biosurfactant increase microorganisms' inherent degrading capacity by lowering cell surface-pollutant interfacial tension. When microbial growth is slow, biosurfactants promote complex chemical biodegradation. Biosurfactants, cationic, anionic, nonionic, and ampholytic surfactants were tested for electroremediation [6]. The biosurfactant β-cyclodextrin degrades TPH at 50%, behind only ampholytic (77%) and anionic (56%) surfactants. Unlike [44], who used SDS, sodium dodecyl benzene sulfonate (LAS), anionic surfactant (Brij 30, nonionic chemical surfactant), Tween 80, and a cyclic lipopeptide biosurfactant to study crude oil biodegradation in an autochthonous microbial community. SDS at onefold CMC was best for crude oil degradation (54.4%), whereas twofold CMC lowered it (51.1%). Brij 30 hurt all concentrations. Biosurfactant increased biodegradation by 0.2 times CMC (53.6%). At higher dosages (equal to CMC, 41.3%), it inhibited the control without biosurfactant (50.5%). Antibacterial agents may interact with pollutants and surfactants, or biosurfactants may hinder solubilization [41].

At low concentrations and severe circumstances, biosurfactants outperform commercial surfactants in biodegradation. Biosurfactants rapidly emulsify oil into microorganism-degradable droplets and increase solubility. Large levels inhibited owing to antibacterial action. Due to their cell disruption capabilities, surfactin and rhamnolipid at 80 mg/L and 240 mg/L inhibit hydrocarbon biodegradation strain formation [45].

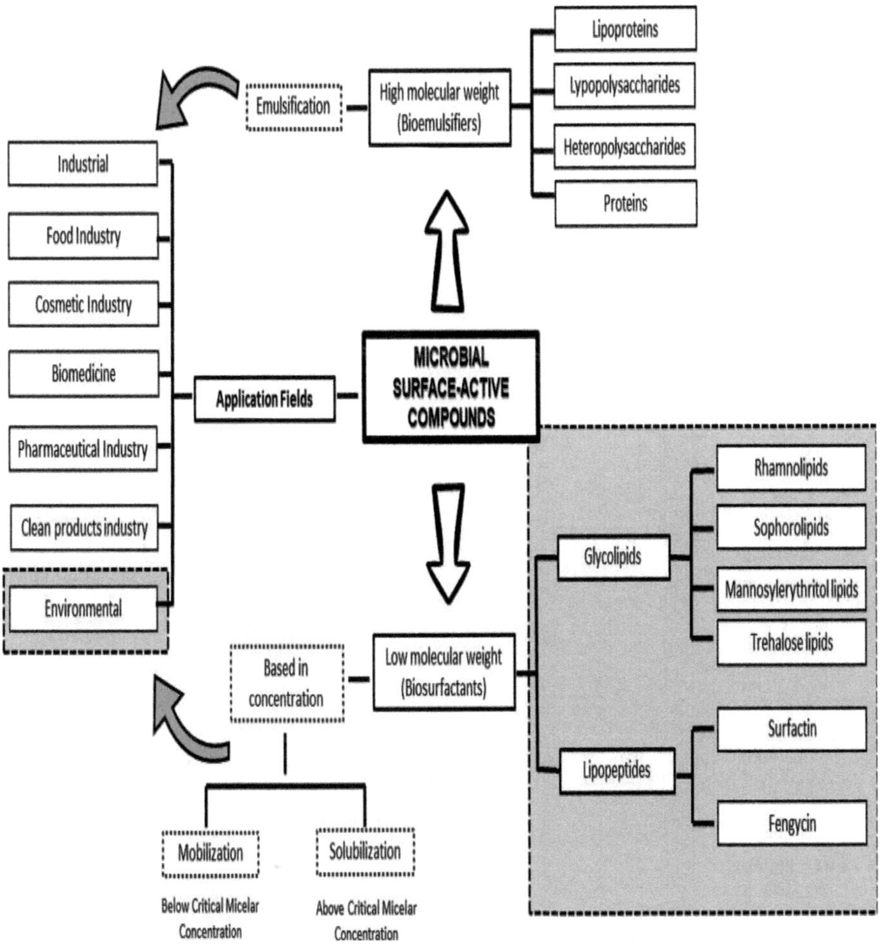


**Fig. 1** Microbial surface-active compound types and characteristics Adopted from [34]

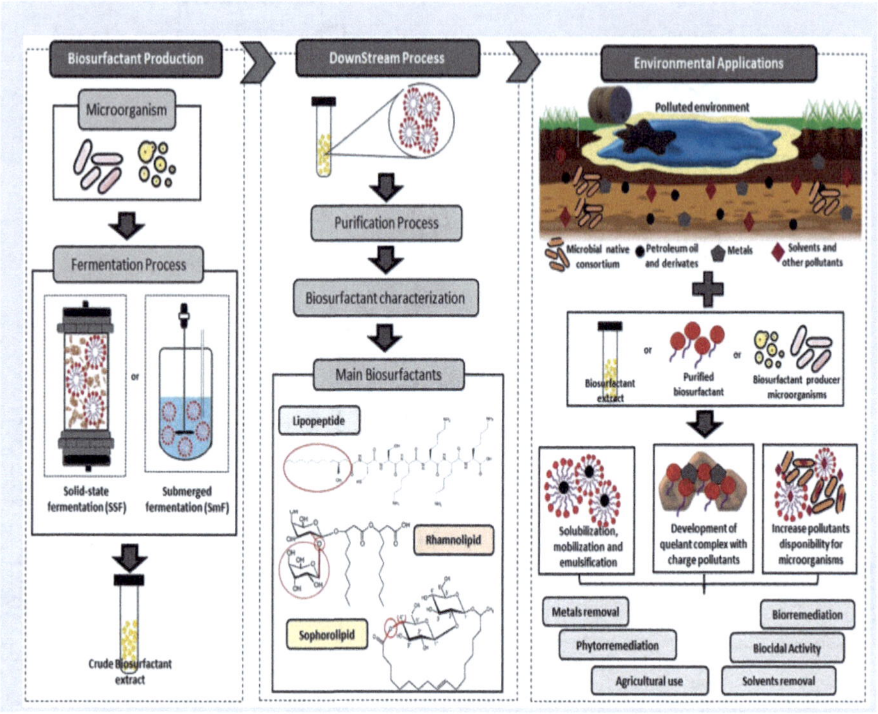


**Fig. 2** Flowchart for biosurfactant production, purification and environment applications (Adopted from Eras-Muñoz et al. [34]

## References


1. Nikolova C, Gutierrez T (2021) Biosurfactants and their applications in the oil and gas industry: current state of knowledge and future perspectives. Front Bioeng Biotechnol 9:626639
2. Brakstad OG, Nordtug T, Throne-Holst M (2015) Biodegradation of dispersed Macondo oil in seawater at low temperature and different oil droplet sizes. Mar Pollut Bull 93(1–2):144–152
3. Zang T, Wu H, Yan B, Zhang Y, Wei C (2021) Enhancement of PAHs biodegradation in biosurfactant/phenol system by increasing the bioavailability of PAHs. Chemosphere 266:128941
4. Farias CBB, Almeida FC, Silva IA, Souza TC, Meira HM, Rita de Cássia F, Sarubbo LA (2021) Production of green surfactants: market prospects. Electron J Biotechnol 51:28–39
5. Ahmad Z, Arshad M, Asghar HN, Sheikh MA, Crowley DE (2016) Isolation, screening and functional characterization of biosurfactant producing bacteria isolated from crude oil contaminated site. Int J Agric Biol 18(3):542–548
6. Li X, Zhao Q, Wang X, Li Y, Zhou Q (2018) Surfactants selectively reallocated the bacterial distribution in soil bioelectrochemical remediation of petroleum hydrocarbons. J Hazard Mater 344:23–32
7. Patel S, Homaei A, Patil S, Daverey A (2019) Microbial biosurfactants for oil spill remediation: pitfalls and potentials. Appl Microbiol Biotechnol 103(1):27–37
8. Mulligan CN (2005) Environmental applications for biosurfactants. Environ Pollut 133(2):183–198

9. Naughton PJ, Marchant R, Naughton V, Banat IM (2019) Microbial biosurfactants: current trends and applications in agricultural and biomedical industries. J Appl Microbiol 127(1):12–28
10. Karlapudi AP, Venkateswarulu TC, Tammineedi J, Kanumuri L, Ravuru BK, Dirisala V, PrabhakarKodali V (2021) Role of biosurfactants in bioremediation of Oil Pollution—a review. Petroleum 7(2):230
11. Mnif I, Ghribi D (2016) Glycolipid biosurfactants: main properties and potential applications in agriculture and food industry. J Sci Food Agric 96(13):4310–4320
12. Wang SH, Tang TWH, Wu E, Wang DW, Wang CF, Liao YD (2020) Inhibition of bacterial adherence to biomaterials by coating antimicrobial peptides with anionic surfactant. Colloids Surf, B 196:111364
13. Fredell DL (2019) Biological properties and applications of cationic surfactants. In: Cationic surfactants (pp 31–60). CRC Press
14. Rosenberg E, Ron EZ (1999) High-and low-molecular-mass microbial surfactants. Appl Microbiol Biotechnol 52:154–162
15. Bages-Estopa S, White DA, Winterburn JB, Webb C, Martin PJ (2018) Production and separation of a trehalolipid biosurfactant. Biochem Eng J 139:85–94
16. Gürkök S, Özdal M (2021) Microbial Biosurfactants: properties, types, and production. Anatol J Biol 2(2):7–12
17. Bollinger A, Thies S, Katzke N, Jaeger KE (2020) The biotechnological potential of marine bacteria in the novel lineage of Pseudomonas pertucinogena. Microb Biotechnol 13(1):19–31
18. Shahaliyan F, Safahieh A, Abyar H (2015) Evaluation of emulsification index in marine bacteria Pseudomonas sp. and Bacillus sp. Arab J Sci Eng 40:1849–1854
19. Chandankere R, Yao J, Cai M, Masakorala K, Jain AK, Choi MM (2014) Properties and characterization of biosurfactant in crude oil biodegradation by bacterium Bacillus methylotrophicus USTBa. Fuel 122:140–148
20. Hassanshahian M, Emtiazi G, Cappello S (2012) Isolation and characterization of crude-oil-degrading bacteria from the Persian Gulf and the Caspian Sea. Mar Pollut Bull 64(1):7–12
21. Labrenz M, Collins MD, Lawson PA, Tindall BJ, Braker G, Hirsch P (1998) Antarctobacter heliothermus gen. nov., sp. nov., a budding bacterium from hypersaline and heliothermal Ekho Lake. Int J Syst Evol Microbiol 48(4):1363–1372
22. Gutierrez T, Morris G, Ellis D, Mulloy B, Aitken MD (2020) Production and characterisation of a marine Halomonas surface-active exopolymer. Appl Microbiol Biotechnol 104:1063–1076
23. Finnerty WR (1992) The genus Rhodococcus identification and classification. Ann Rev Microbiol 46:193–218
24. Wang W, Zhong R, Shan D, Shao Z (2014) Indigenous oil-degrading bacteria in crude oil-contaminated seawater of the Yellow sea, China. Appl Microbiol Biotechnol 98:7253–7269
25. Peng F, Liu Z, Wang L, Shao Z (2007) An oil-degrading bacterium: Rhodococcus erythropolis strain 3C–9 and its biosurfactants. J Appl Microbiol 102(6):1603–1611
26. Cai Q, Zhang B, Chen B, Zhu Z, Lin W, Cao T (2014) Screening of biosurfactant producers from petroleum hydrocarbon contaminated sources in cold marine environments. Mar Pollut Bull 86(1–2):402–410
27. Chikkanna A, Ghosh D, Kishore A (2018) Expression and characterization of a potential exopolysaccharide from a newly isolated halophilic thermotolerant bacteria Halomonas nitroreducens strain WB1. PeerJ 6:e4684
28. Béjar V, Llamas I, Calvo C, Quesada E (1998) Characterization of exopolysaccharides produced by 19 halophilic strains of the species Halomonas eurihalina. J Biotechnol 61(2):135–141
29. Amjres H, Béjar V, Quesada E, Carranza D, Abrini J, Sinquin C, Llamas I (2015) Characterization of haloglycan, an exopolysaccharide produced by Halomonas stenophila HK30. Int J Biol Macromol 72:117–124
30. Ruiz-Ruiz C, Srivastava GK, Carranza D, Mata JA, Llamas I, Santamaría M, Molina IJ (2011) An exopolysaccharide produced by the novel halophilic bacterium Halomonas stenophila strain B100 selectively induces apoptosis in human T leukaemia cells. Appl Microbiol Biotechnol 89:345–355

31. Holmström C, Kjelleberg S (1999) Marine Pseudoalteromonas species are associated with higher organisms and produce biologically active extracellular agents. FEMS Microbiol Ecol 30(4):285–293
32. Liu JF, Mbadinga SM, Yang SZ, Gu JD, Mu BZ (2015) Chemical structure, property and potential applications of biosurfactants produced by Bacillus subtilis in petroleum recovery and spill mitigation. Int J Mol Sci 16(3):4814–4837
33. Gutiérrez T, Mulloy B, Bavington C, Black K, Green DH (2007) Partial purification and chemical characterization of a glycoprotein (putative hydrocolloid) emulsifier produced by a marine bacterium Antarctobacter. Appl Microbiol Biotechnol 76:1017–1026
34. Eras-Muñoz E, Farré A, Sánchez A, Font X, Gea T (2022) Microbial biosurfactants: a review of recent environmental applications. Bioengineered 13(5):12365–12391
35. Machado TS, Decesaro A, Cappellaro AC, Machado BS, van Schaik Reginato K, Reinehr CO, Colla LM (2020) Effects of homemade biosurfactant from Bacillus methylotrophicus on bioremediation efficiency of a clay soil contaminated with diesel oil. Ecotoxicol Environ Saf 201:110798
36. Martins PC, Martins VG (2018) Biosurfactant production from industrial wastes with potential remove of insoluble paint. Int Biodeterior Biodegradation 127:10–16
37. Naeem U, Qazi MA (2020) Leading edges in bioremediation technologies for removal of petroleum hydrocarbons. Environ Sci Pollut Res 27(22):27370–27382
38. Prakash AA, Prabhu NS, Rajasekar A, Parthipan P, AlSalhi MS, Devanesan S, Govarthanan M (2021) Bio-electrokinetic remediation of crude oil contaminated soil enhanced by bacterial biosurfactant. J Hazard Mater 405:124061
39. Imam A, Kanaujia PK, Ray A, Suman SK (2021) Removal of petroleum contaminants through bioremediation with integrated concepts of resource recovery: a review. Indian J Microbiol 61:250–261
40. Wei Z, Wang JJ, Gaston LA, Li J, Fultz LM, DeLaune RD, Dodla SK (2020) Remediation of crude oil-contaminated coastal marsh soil: Integrated effect of biochar, rhamnolipid biosurfactant and nitrogen application. J Hazard Mater 396:122595
41. Patowary R, Patowary K, Kalita MC, Deka S (2018) Application of biosurfactant for enhancement of bioremediation process of crude oil contaminated soil. Int Biodeterior Biodegradation 129:50–60
42. Chaprão MJ, Ferreira IN, Correa PF, Rufino RD, Luna JM, Silva EJ, Sarubbo LA (2015) Application of bacterial and yeast biosurfactants for enhanced removal and biodegradation of motor oil from contaminated sand. Electron J Biotechnol 18(6):471–479
43. Ayed HB, Jemil N, Maalej H, Bayoudh A, Hmidet N, Nasri M (2015) Enhancement of solubilization and biodegradation of diesel oil by biosurfactant from Bacillus amyloliquefaciens An6. Int Biodeterior Biodegradation 99:8–14
44. Tian W, Yao J, Liu R, Zhu M, Wang F, Wu X, Liu H (2016) Effect of natural and synthetic surfactants on crude oil biodegradation by indigenous strains. Ecotoxicol Environ Saf 129:171–179
45. Sajna KV, Sukumaran RK, Gottumukkala LD, Pandcy A (2015) Crude oil biodegradation aided by biosurfactants from Pseudozyma sp. NII 08165 or its culture broth. Biores Technol 191:133–139

# Sugar Based Gelator for Oil Spill Treatment

**Kenneth Chinekwu Ugwuoke, Isaac John Umaru, Kayode Adebisi Arowora, and Kerenhappuch Isaac Umaru Moses Adondua Abah**

**Abstract** Oil spills pose significant threats to the environment, demanding effective cleanup strategies. This chapter investigates the utilization of sugar-based gelators, specifically low molecular weight carbohydrates, for oil spill treatment. Through a detailed classification, we emphasize the structural features and gelation characteristics of D-glucose derivatives and disaccharide derivatives as efficient low molecular weight gelators (LMWGs). The natural origin, biodegradability, and stabilizing gel network formation ability of sugar-based gelators make them invaluable in mechanically addressing oil spills. This study employs a supramolecular gelation approach, employing glucose-derived organogelators and hydrogelators, showcasing their remarkable efficiency in transforming spilled oil into a manageable gel-like state. This process enables straightforward removal and recovery through heating, providing a robust cleanup method. Our research contributes to a deeper understanding of the specific advantages offered by D-glucose and disaccharide derivatives as LMWGs in oil spill treatment. By highlighting their structural attributes and gelation efficiency, this chapter underscores the potential of sugar-based gelators as a crucial component in addressing environmental challenges.

## 1 Introduction

A gel is a viscoelastic component made up of solvents and gelator particles with a low atomic weight [1]. Their characteristics range from being soft and weak to being firm and durable, indicating that they are in a semi-solid condition [1, 2]. Gels are defined as cross-linked systems that are very dilute and do not exhibit flow at steady state, even if the liquid phase may diffuse through them [3]. From a phenomenological standpoint, a gel is defined as a material that is either soft, solid, or has the consistency of a solid and has two or more components, with a liquid component being present in

K. C. Ugwuoke · I. J. Umaru (✉) · K. A. Arowora · K. I. U. M. A. Abah
Department of Biochemistry, Faculty of Pure and Applied Sciences, Federal University Wukari, Wukari, Nigeria
e-mail: umaruisaac@gmail.com

© The Author(s), under exclusive license to Springer Nature Singapore Pte Ltd. 2025 
S. A. Bhawani et al. (eds.), *Biobased Materials and Their Composites for Oil Spill Treatment*, Composites Science and Technology,
https://doi.org/10.1007/978-981-96-6049-0_11

significant amounts [4]. A three-dimensional cross-linked network provides structure and sticky stick (tack) in gels, which cause them to behave like solids despite being mostly liquid by mass [1]. Things that can produce gels, called gelators, make the process of gelation easier [5].

Low Molecular Weight Gelators (LMWGs) are among the many types of gelators that can be defined by their chemical makeup and the characteristics they display [5]. Supramolecular gels, or long, anisotropic structures (LMWGs) are formed when tiny molecules self-assemble into structures, most often fibers [6–10]. The solvent is immobilized by a network of entangled or cross-linked fibers at greater concentrations, which is achieved by surface tension and capillary forces [6, 7]. The cross-linking of supramolecular gels can be reversed, for instance, by heating, which is a distinguishing feature compared to covalently cross-linked polymer gels [11].

A subclass of carbohydrate-derived LMWGs known as sugar-based gelators can, in particular, create gels in water [12]. With hydrophilic and hydrophobic components, sugar-based gelators self-assemble, forming a gel network vital for many applications such as in food goods, medicines, and cosmetics [13]. The use of sugar-based gelators in the cleanup of oil spills is explored in detail in this chapter.

There has been a lot of focus on low molecular weight carbohydrates as gelators recently. Non-covalent interactions like as hydrogen bonding, hydrophobic contacts, π–π stacking, CH–π interactions, and van der Waals forces, among others, can be effectively used in the creation of compound structures using LMWGs to cause gelation [14]. The flexible and intriguing features of supramolecular gels generated by LMWGs, particularly hydrogels, make them ideal for use in tissue engineering systems and drug delivery vehicles, among other biological applications [15–17]. Porous materials entrapping solvent molecules inside a solid gel matrix generate supramolecular gels with distinctive morphologies, and LMWGs are one example of this [18]. The architectures of LMWGs include numerous classes of small molecules and a wide variety of functions [19]. Aggregates can take several shapes, including two-dimensional objects like platelets or one-dimensional fibers, rods, ribbons, or nanotubes. Gels are made when these particles immobilize the solvent component by further organizing into three-dimensional networks. So, LMWG can be thought of as porous materials with solvent molecules trapped in the gaps between the pores of a [20] solid, stiff gel matrix.

Notable features of LMWGs include their potential for degradability and flammability, ease of manufacture and modification, and their ibility to dissolve into a solvent of similar thickness, all of which contribute to their wide range of potential uses [21]. Numerous studies have investigated LMWGs, looking at their structure–property relationship and its uses in areas such as controlled release, bioactivity support, micro and nano material design, and oil recovery [22] (Fig. 1).

Because of the grave dangers that oil spills pose to ecosystems, human health, and economic stability, their cleanup is an urgent matter. Cleaning up an oil spill is a difficult and time-consuming task that calls for a variety of approaches depending on variables such the oil type, sea temperature, and coastline features [24, 25]. In mechanical cleanup, the gelator approach offers advantages, especially in difficult situations like turbulent waters or when dealing with specific types of oil [26]. This method

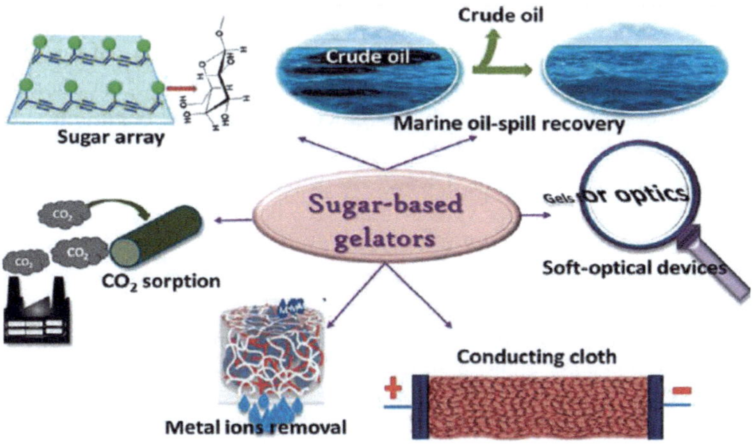


**Fig. 1** Sugar-based gelators for various applications [23]

involves using gelators to turn spilled oil into a gel-like state. Natural, biodegradable, and capable of creating gels, sugar-based gelators.

## 2 Classification of Sugar-Based Gelators

Sugar-based gelators can be categorized based on various criteria, such as their chemical composition, origin, or mechanism of gelation. A comprehensive classification based on their chemical structure is presented below.

### *2.1 Structural Features and Gelation Characteristics of Sugar-Based Gelators*

The most abundant renewable resource in nature are carbohydrates, which, along with proteins and nucleic acids, are important biomolecules. Carbohydrates are involved in many biological processes and disease-related occurrences, and they are characterized by their rich stereochemistry and extensive structural variety. Because they are probably biocompatible, sugar-based gelators have a lot of potential as novel materials [27, 28]. Several classes of effective hydrogelators and organogelators with low molecular weight have been developed, with Wang's group and others taking advantage of the availability of monosaccharides and disaccharides in nature.

**D-Glucose Derivatives as LMWGs**: Wang and colleagues have focused their research on developing, synthesizing, and characterizing LMWGs (low molecular weight gelators) from D-glucose. In recent years, a wide variety of monosaccharide derivatives have been found to be efficient hydrogelators and organogelators [29–42]. Various LMWGs have been synthesized by selectively functionalizing 4,6-benzylidene acetal-protected α-methyl D-glucose derivatives, beginning with easily accessible D-glucose. Functionalized derivatives at various sites have had their gelation properties studied in detail. Diester derivative 2 and monoester derivatives 3 and 4 are among the acyl derivatives that have been produced and reviewed. At 5 mg/mL, diester 2 h gelated in EtOH:H2O (v/v 1:1) [32], and at 7 mg/mL, 3-ester derivative 4d formed a hydrogel, despite the fact that several saturated alkyl chain esters did not work as gelators. The ability to gel was demonstrated for a variety of solvents, including hexane and water, by short-chain esters containing terminal alkyne functional groups, such as 2a-c, 3a-c, and 4a-c (Fig. 2).

**Disaccharide Derivatives**: A number of disaccharide triazole derivatives were produced and tested for their effectiveness as LMWGs in an effort to comprehend the correlations between sugar structure and gelation [41]. Thirteen peracetyl maltosyl triazole derivatives and nine peracetyl lactosyl triazole derivatives were produced during the synthesis. The majority of lactosyl derivatives did not gel, while compound 59 did so in a 1:1 mixture of EtOH and water, as well as a 1:1 mixture of DMSO and water. In contrast, maltose-based disaccharides were excellent gelators. For example, at concentrations between 2.0 and 5.0 mg/mL, compound 60d effectively formed a gel in water-based DMSO or ethanol combinations. At doses between 2.0 and 6.7 mg/

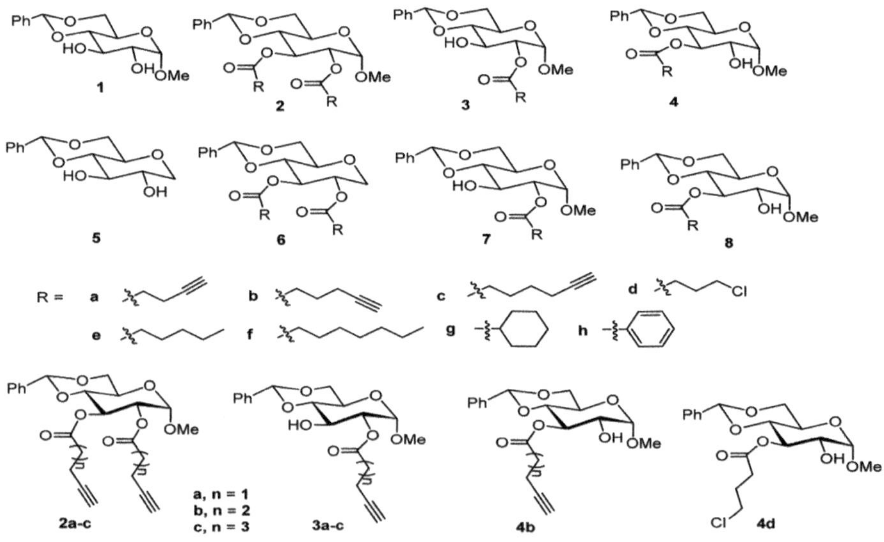


**Fig. 2** General structures of several classes of glucose ester derivatives as LMWGs [31, 32], n represent the number of repeating $CH_2$ [43]

mL, another derivative, 60e, showed stable gel formation in ethanol and water combinations [41]. The structural patterns of the sugar groups may be associated with the discrepancies in gelation capabilities between the two types of disaccharides [41].

Two amphiphilic LMWGs based on thiolactose were designed and synthesized by Cano et al. [44]. The two amphiphiles were identical save for the length of the ethylene glycol linker; they were both constructed from a di-lauroyl-L-tartaric acid scaffold. At a concentration of 0.1% w/v, the hydrogel was produced by the shorter linker variation 61a, while the lactose-containing amphiphiles created colloidal systems in water. This highlights the substantial effect of the length of the ethylene glycol linker on the chiral self-assembly of these molecules, since circular dichroism of the self-assembled aggregates showed a chiral inversion for amphiphile 61b as opposed to 61a. In both instances, agglutination was seen when the amphiphiles were tested for their binding capabilities to peanut agglutinin (Pna) lectin. Insights into the rational design of nanostructures exposing specific sugar ligands were gained from this study, which investigated structural aspects impacting amphiphilic self-assembly.

## 3 Application of Sugar-Based Gelators in the Treatment of Oil Spills

Marine oil spills are a major problem for the environment because they can happen for a variety of reasons, including the intentional and unintentional release of petrochemicals and crude oil into water. Pollution, devastating effects on marine ecosystems, and the discharge of non-renewable fossil fuels are among the disastrous outcomes of the more than a thousand oil spills reported each year, according to statistics. Absorption, dispersion, combustion, bioremediation, fertilizers, and other strategies are employed to address oil spills [7–10]. The extraordinary efficacy of supramolecular gelation distinguishes it from the other methods. By causing the oil on the water's surface to gel, this procedure makes it possible to heat the gel and extract the oil, making its removal much easier. In addition to making oil recovery easier, these gelation technologies are thermo-reversible and non-destructive, so the gelator itself may be recycled and reused [11–15].

It has been observed that in many gelation methods, the gelator is dissolved into oil through heating and then cooled to produce the gel. There are several methods for creating gels, such as dissolving the gelator in a solvent that dissolves oil or water, or just applying oil. Because of non-covalent interactions that can be used to reverse gelation, the gels show reactivity to environmental factors like light, temperature, and pH.

Organic methods, frequently utilizing compounds obtained from petroleum, have been employed to synthesis a plethora of gelators, drawing from sugars and sugar derivatives such as galactose, mannitol, xylitol, and *N*-acetyl glucosamine. These gelators show rapid initial gelation, usually in seconds, but it takes a few minutes to more than an hour for the gel to harden. For effective oil and gelator recovery, the gel's mechanical qualities are critical for efficient removal from the water surface and subsequent handling. In many instances, the gelator can be reused numerous times, and oil recovery often requires distillation [45, 46]. Some oils demand high temperatures for evaporation, hence the gelator's reusability depends on how stable it is at different temperatures.

To dissolve the gelator before application, many of the gelators listed below need a cosolvent such THF, toluene, or ethanol. In order to dissolve the gelator solution thoroughly, it is often necessary to heat it. On the other hand, there are certain environmental concerns and issues with temperature management that make using a cosolvent problematic in practical settings. An improved alternative that does away with the requirement for a cosolvent is powdered gelators. While this method improves safety and tackles environmental concerns, it slows down the diffusion of the gelator in the oil phase, which causes uneven distribution. However, it is still preferable than using a cosolvent (Fig. 3).

The organogelator that Vibhute et al. [48] created from D-Glucose might be used to create a gel by mixing it with seawater and then applying it as a fine powder to benzene, diesel, and crude oil. At a 1% (w/v) concentration, the gelator, alkyl 4,6-*O*-benzylidene-glycopyranoside, showed encouraging results, creating a firm gel in just one hour that could be easily scraped off the water's surface. Following this, the oil may be extracted by distillation, and the gelator could be recycled for other uses [48].

The cosolvent problem was solved by using a two-component deep eutectic solvent (DES) system, which combines a solid gelator with a solid co-solvent to create a mixture that has a lower melting point than the individual components. An efficient DES was prepared by mixing lauric acid and *N*-methylacetamide; it was then

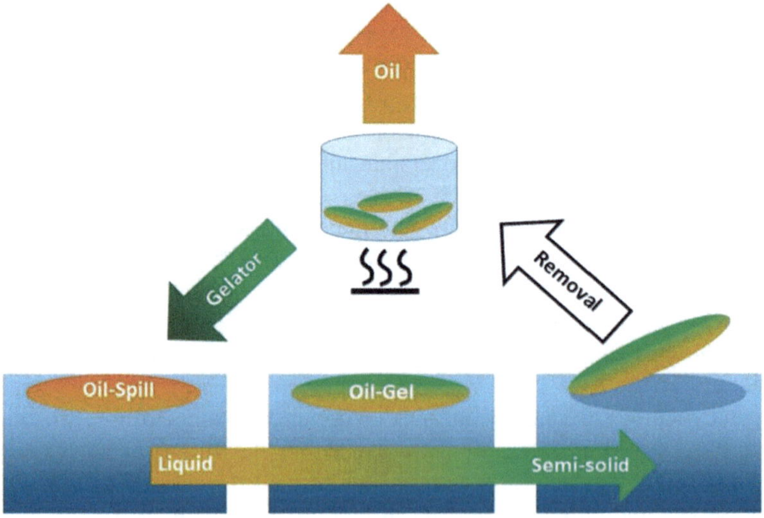


**Fig. 3** Oil recovery using gelators [47]

used to the problem of oil gelation. By reducing the potential toxicity of conventional cosolvents, this method offers a greener alternative [49].

Researchers Vibhute et al. [50] created phase-selective organogelators using glucose as their starting material. Organogelators are a viable option for recovering oilspills because they can operate as solids independently of a carrier solvent. The newly proposed technique overcomes a significant constraint of low-molecular-mass organogelators (LMOGs) that has been previously proposed for maritime oil-spill remediation: the need for a carrier solvent. Their creative approach utilizes a steady LMOG, offering a practical mechanism for efficient oil-spill cleanup.

Amphiphilic diols 6 and 7 were produced in a related investigation by partially protecting the hydroxyl groups of mannitol as ketals. These diols showed promising gelation characteristics with a range of oils and nonpolar solvents. The importance of vicinal diol functionality and ketal protecting groups was highlighted in structural activity correlation research. The importance of hydrogen bonding in the gelation process was shown by spectroscopic investigations. A three-dimensional fibrous network is created when gelators self-assemble by hydrogen bonding, according to the hypothesized method (Fig. 4).

The synthesis of 3,4,5-tri-*O*-alkyl substituted benzohydrazide-based *N*-glycosylamines was described in detail by Soundarajan et al. [45]. These chemicals showed that they could dissolve organic solvents, petroleum, and coconut oil in water, forming phase-selective gels. They were also tested for their ability to remove water-soluble dyes. The synthesis of a stable gel in both polar and non-polar solvents was

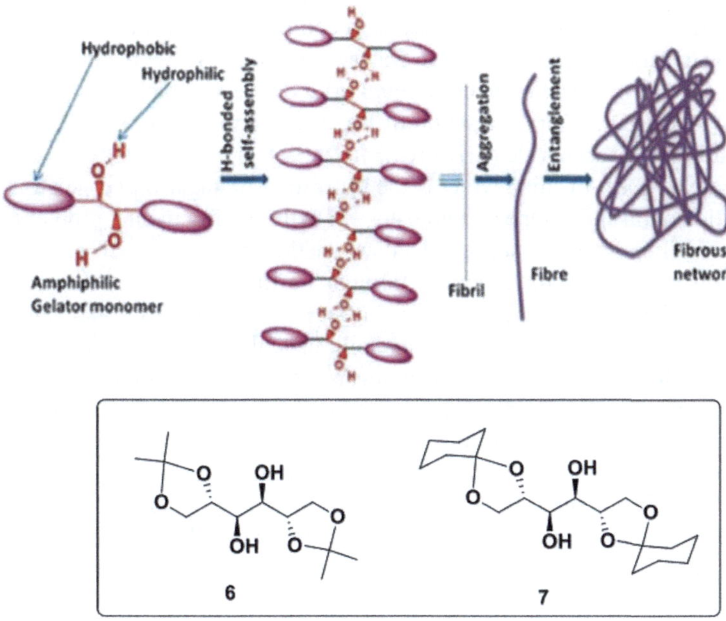


**Fig. 4** Proposed mechanism of gelation for diols 6 and 7 [51]

also included in the study for a 4-*O*-octyl substituted oxadiazole *N*-glycosylamine. The production of gels was facilitated by weak contacts such hydrogen bonding, π–π interactions, and dipole–dipole interactions, which were vital in the self-assembly of sugar derivatives based on oxadiazole. Gelation experiments demonstrated the synthesized compounds' potential as phase-selective gelators for a variety of uses, including crude oil spill recovery, and characterization using H and C NMR and mass spectral analysis validated the compounds.

It was determined that the gelators could recover crude oil spills since they formed gels efficiently, especially in less time. The driving forces behind the creation of the gel were hydrogen bonding and π–π stacking interactions, as revealed by FT-IR and 1H NMR investigations. When tested with saltwater, the synthesized gelator demonstrated promise in entrapping watery synthetic dyes and recovering crude oil, indicating its potential for use in water treatment and other fields [46].

# 4 Conclusion

A major step forward in materials science and green technology has been the investigation of sugar-based gelators. An in-depth comprehension of the adaptability and potential of these gelators is provided by their different classification according to chemical structure, source, and gelation mechanisms. The exceptional qualities of Low Molecular Weight Gelators (LMWGs), especially those produced from sugars, render them indispensable in a multitude of sectors. Opportunities for their extensive use arise from sugar-based gelators' exceptional self-assembly and gel network formation capabilities, particularly in watery conditions. These materials have great promise for use in tissue engineering and drug delivery vehicles, as demonstrated by the synthesis and characterisation of D-glucose derivatives as LMWGs and the investigation of disaccharide derivatives.

A notable aspect of this review is the utilization of sugar-based gelators for the remediation of oil spills. One potential answer to the environmental problems caused by marine oil spills is the supramolecular gelation method, which efficiently gels surface oil and allows for recovery through heating. Another realistic aspect of powdered gelators is their potential as a better alternative to co-solvents, which would increase environmental safety and eliminate the need for them.

Moving ahead, new discoveries in sugar-based gelators could lead to groundbreaking approaches to cleaning up oil spills and other environmental disasters. The gelators are designed to be reusable and recyclable, which helps promote sustainable practices and is in line with the worldwide movement towards eco-friendly technology. To sum up, sugar-based gelators are incredibly adaptable and have great promise in a wide range of industries, including biomedicine and environmental cleanup. If we want to find all the uses for these materials and make sure they can withstand the test of time, we need to keep digging into them.

## References


1. Khademhosseini A, Demirci U (2016) Gels handbook: fundamentals, properties and applications. World Scientific Pub Co Inc.
2. Seiffert S (ed) (2015) Supramolecular polymer networks and gels. Springer. ASIN B00VR5CMW6
3. Ferry JD (1980) Viscoelastic properties of polymers. Wiley, New York. ISBN 0471048941
4. Almdal K, Dyre J, Hvidt S, Kramer O (1993) Towards a phenomenological definition of the term 'gel.' Polym Gels Netw 1(1):5–17. https://doi.org/10.1016/0966-7822(93)90020-I
5. Harper D (2013) Online etymology dictionary: gel. Online etymology dictionary. Retrieved 09 Dec 2013
6. Estroff LA, Hamilton AD (2004) Water gelation by small organic molecules. Chem Rev 104:1201–1218
7. Terech P, Weiss RG (1997) Low molecular mass gelators of organic liquids and the properties of their gels. Chem Rev 97:3133–3160
8. Du X, Zhou J, Shi J, Xu B (2015) Supramolecular hydrogelators and hydrogels: from soft matter to molecular biomaterials. Chem Rev 115:13165–13307
9. Weiss RG (2014) The past, present, and future of molecular gels. What is the status of the field, and where is it going? J Am Chem Soc 136:7519–7530
10. Sangeetha NM, Maitra U (2005) Supramolecular gels: functions and uses. Chem Soc Rev 34:821–836
11. Jones CD, Steed JW (2016) Gels with sense: supramolecular materials that respond to heat, light and sound. Chem Soc Rev 45:6546–6596
12. Dhinakaran MK, Mohan Das T (2012) Org Biomol Chem 10:2077–2083
13. Osada Y, Kajiwara K (2001) Gels handbook: applications. Academic Press
14. (a) Sangeetha NM, Maitra U (2005). Chem Soc Rev 34:821–836; (b) Zhao F, Ma ML, Xu B (2009). Chem Soc Rev 38:883–891; (c) Kumar KK, Elango M, Subramanian V, Mohan Das T (2009). NewJ Chem 33:1570–1577; (d) Ajayaghosh A, Praveen VK, Vijayakumar C, George SJ (2007). Ang Chem Int 46:6260–6265
15. (a) Hirst AR, Escuder B, Miravet JF, Smith DK (2008). Angew Chem Int Ed 47:8002–8018; (b) Koshi Y, Nakata E, Yamane H, Hamachi I (2006). J Am Chem Soc 128:10413–10422; (c) Matsumoto S, Yamaguchi S, Ueno S, Komatsu H, Ikeda M, Ishizuka K, Iko Y, Tabata KV, Aoki H, Ito S, Noji H, Hamachi I (2008). Chem Eur J 14:3977–3986
16. (a) Yang Z, Liang G, Guo Z, Guo Z, Xu B (2007). Angew Chem Int Ed 46:8216–8219; (b) Yang C, Li D, Liu Z, Hong G, Zhang J, Kong D, Yang Z (2012). J Phys Chem B 116:633–638
17. (a) Ray S, Das AK, Banerjee A (2007). Chem Mater 19:1633–1639; (b) Wang H, Yang Z (2012). Nanoscale 4:5259–5267
18. (a) Sangeetha N, Maitra U (2005). Chem Soc Rev 34:821–836; (b) de Loos M, Feringa BL, van Esch JH (2005). Eur J Org Chem 3616–3631
19. (a) Peters GM, Davis JT (2016). Chem Soc Rev 45:3188−3206; (b) Babu SS, Praveen VK, Ajayaghosh A (2014). Chem Rev 114:1973−2129; (c) Estroff LA, Hamilton AD (2004). Chem Rev 104:1201−1217; (d) Lau KH (2014). A Biomater Sci 2:627−633
20. Terech PR, Weiss G (1997) Chem Rev 97:3133–3159
21. Yang Y, Suzuki M, Owa S, Hanabusa K (2007) J Am Chem Soc 129:581–587
22. (a) Khatua D, Dey J (2005). Langmuir 21:109–114; (b) Xue PC, Lu R, Li D, Jin MM, Tan CH, Bao CY, Wang ZM, Zhao YY (2001). Langmuir 20:11234–11239; (c) Yang YG, Suzuki M, Shirai H, Kurose A, Hanabusa K (2005). Chem Commun 2032–2034
23. Ren F, Chen F, Zhou J, Shen H, Su, Zeng H (2016). Langmuir 32(50):13510–13516
24. Lingering Lessons of the Exxon Valdez Oil Spill. Commondreams.org. 2004–03–22. Archived from the original on June 13, 2010. Retrieved 27 Aug 2012
25. NOAA Ocean Media Center (2010–03–16) Hindsight and foresight, 20 years after the Exxon Valdez spill. NOAA. Retrieved 30 April 2010
26. Broekema W (2015) Crisis-induced learning and issue politicization in the EU. Public Admin 94(2):381–398. https://doi.org/10.1111/padm.12170

27. Datta S, Bhattacharya S (2015) Multifarious facets of sugar-derived molecular gels: molecular features, mechanisms of self-assembly and emerging applications. Chem Soc Rev 44:5596–5637
28. Soundarajan K, Periyasamy R, Mohan Das T (2016) Design and synthesis of sugar-benzohydrazides: low molecular weight or-ganogelators. RSC Adv 6:81838–81846
29. Wang G, Cheuk S, Williams K, Sharma V, Dakessian L, Thorton Z (2006) Synthesis and characterization of monosaccharide lipids as novel hydrogelators. Carbohydr Res 341:705–716
30. Nie X, Wang G (2006) Synthesis and self-assembling properties of diacetylene-containing glycolipids. J Org Chem 71:4734–4741
31. Wang G, Yang H, Cheuk S, Coleman S (2011) Synthesis and self-assembly of 1-deoxyglucose derivatives as low molecular weight organogelators. Beilstein J Org Chem 7:234–242
32. Cheuk S, Stevens ED, Wang G (2009) Synthesis and structural analysis of a series of D-glucose derivatives as low molecular weight gelators. Carbohydr Res 344:417–425
33. Goyal N, Cheuk S, Wang G (2010) Synthesis and characterization of d-glucosamine-derived low molecular weight gelators. Tetrahedron 66:5962–5971
34. Wang G, Cheuk S, Yang H, Goyal N, Reddy PVN, Hopkinson B (2009) Synthesis and characterization of monosaccharide-derived carbamates as low-molecular-weight gelators. Langmuir 25:8696–8705
35. Goyal N, Mangunuru HPR, Parikh B, Shrestha S, Wang G (2014) Synthesis and characterization of pH responsive D-glucosamine based molecular gelators. Beilstein J Org Chem 10:3111–3121
36. Mangunuru HPR, Yerabolu JR, Wang G (2015) Synthesis and study of *N*-acetyl D-glucosamine triazole derivatives as effective low molecular weight gelators. Tetrahedron Lett 56:3361–3364
37. Mangunuru HPR, Yang H, Wang G (2013) Synthesis of peptoid based small molecular gelators by a multiple component reaction. Chem Commun 49:4489–4491
38. Wang G, Goyal N, Mangunuru HPR, Yang H, Cheuk S, Reddy PVN (2015) Preparation and self-assembly study of amphiphilic and bispolar diacetylene-containing glycolipids. J Org Chem 80:733–743
39. Mangunuru HPR, Yerabolu JR, Liu D, Wang G (2015) Synthesis of a series of glucosyl triazole derivatives and their self-assembling properties. Tetrahedron Lett 56:82–85
40. Wang G, Chen A, Mangunuru HPR, Yerabolu JR (2017) Synthesis and characterization of amide linked triazolyl glycolipids as molecular hydrogelators and organo-gelators. RSC Adv 7:40887–40895
41. Okafor IS, Wang G (2017) Synthesis and gelation property of a series of disaccharide triazole derivatives. Carbohydr Res 451:81–94
42. Chen A, Adhikari SB, Mays K, Wang G (2017) Synthesis and study of molecular assemblies formed by 4,6-*O*-(2-phenylethylidene)-functionalized D-glucosamine derivatives. Langmuir 33:8076–8089
43. Joedian M, Jonathan B, Kristen B, Guijun W (2021) Recently developed carbohydrate based gelators and their applications. Gels 7:24. https://doi.org/10.3390/gels7010024
44. Cano ME, Di Chenna PH, Lesur D, Wolosiuk A, Kovensky J, Uhrig ML (2017) Chirality inversion, supramolecular hydrogelation and lectin binding of two thiolactose amphiphiles con-structed on a di-lauroyl-l-tartaric acid scaffold. New J Chem 41:14754–14765
45. Mukherjee S, Mukhopadhyay B (2013) Phase selective carbohydrate gelator. RSC Adv 2:2270–2273
46. Narayana C, Upadhyay RK, Chaturvedi R, Sagar R (2017) A versatile carbohydrate based gelator for oil water separation, nanoparticle synthesis and dye removal. New J Chem 41:2261
47. Doshi B, Sillanpää M, Kalliola S (2018) A review of bio-based materials for oil spill treatment. Water Res 135(15):262–277
48. Vibhute AM, Muvvala V, Sureshan KM (2016) A sugar-based gelator for marine oil-spill recovery. Angew Chem Int Ed 55:7782–7785
49. Cui Y, Li M-C, Wu Q, Pojman JA, Daniel G (2017) Kuroda Synthesis-free phase-selective gelator for oil spill remediation. ACS Appl Mater Interfaces 9:33549–33553

50. Vibhute AM, Muvvala V, Sureshan KM (2016) A sugar-based gelator for marine oil-spill recovery. Angew Chem 128:7913–7916
51. Rabecca Jenifer V, Akshay MS, Dhivahar J, Das TM (2021) Emerging perspectives of sugar-based gelators of diverse applications. Trends Carbohydr Res 13(2):15–34

# Cellulose Nanocrystals for Oil Spill Treatment

**Atif Husain and Benjamin Siddiqui**

**Abstract** The past few decades have seen intense use of biobased materials for versatile applications in various haphazard catastrophes, one of which is oil spill treatment. Falling in the same list with exclusive properties and diverse applications is cellulose nanocrystals (CNCs). The employment of cellulose nanocrystals due to its potential ability of oil spill treatment paves way for the development of newer strategies and updating the present strategies. An urgent need for the implementation of these biobased materials, including cellulose nanocrystals, is very essential in the present scenario to tackle the drastic consequences of oil spill or leakage. These remedial measures and the persistent challenges in their implementation also need to be equally addressed. This chapter presents a brief understanding of cellulose nanocrystals as one of the leading biobased materials by exploring their morphology and unique properties. The treatment, present strategies, and remedial measures need to be strictly followed to delimitate the global phenomenon of oil spill leakage and its treatment is also equally addressed.

**Keywords** Cellulose · Cellulose nanocrystals · Cellulose nanofibrils · Hydrolysis · Oil spill · Sustainability

## 1 Introduction

Oil leaks and spills harm the environment for a long time, especially the neighbouring marine ecosystems. The Deepwater Horizon oil spill in 2010 is among the most well-known in history. The Gulf of Mexico saw the incident, which is regarded as the biggest marine oil spill. Experimental products, some of which by Environmental Protection Agency (EPA) had not approved for use to mitigate oil spills, were widely used to address the leak and safeguard local flora and fauna [1].

A. Husain (✉) · B. Siddiqui
Department of Chemistry, Integral University, Dasauli, Lucknow, India
e-mail: Kinghusain111@gmail.com

© The Author(s), under exclusive license to Springer Nature Singapore Pte Ltd. 2025 
S. A. Bhawani et al. (eds.), *Biobased Materials and Their Composites for Oil Spill Treatment*, Composites Science and Technology,
https://doi.org/10.1007/978-981-96-6049-0_12

Major occupational, environmental, and public health catastrophes are frequently caused by oil spills [2]. Crude oil spills could have a harmful impact upon that ecosystem's flora and fauna if adequate spill remediation techniques are not employed [3]. The infusion of surface-active chemicals or dispersants into spilt oil to split it up into small droplets is frequently useful in treating oil spills [4]. Dispersants and lipophilic fertilisers are examples of surface-active substances that can expedite the biodegradation process at the oil–water interface. Surfactants are frequently used with organic solvents such glycol and light petroleum distillates in dispersants used it to clean up oil spills [5].

On the other hand, Cellulose nanocrystals with various exclusive properties due to their structural morphology and variability have proven to be a leading remedial tool for various consequences. The materials world has given cellulose nanocrystals (CNs) a great deal of attention, and this interest does not appear to be fading. These biopolymeric structures seem deserving of such attention due to their abundance as well as their unrivalled fundamental physical and chemical characteristics. They also deserve attention due to their intrinsic renewability and sustainability. They have been the subject of a variety of research efforts as reinforcement material in nanocomposites for their low cost, affordability, recyclability, light weight, nanoscale size, and unique shape. In reality, cellulosic materials' macroscopic fibers' fundamental polymeric constitutive motifs are CNs, and these fibers' sheer mass dwarfs those of any other known naturally occurring or artificial biomaterial. Cellulose nanocrystals are frequently referred to as nanocrystals, whisker, nanoparticles, particles, microcrystallites, and nanofibers.

This chapter will present a brief view on the disastrous ill-effects of oil spill and leakage accidents globally along with the present strategies to tackle this looming catastrophe already present and are being discovered for oil spill treatment. Implementation of cellulose nanocrystals as a leading biobased material with its unique properties, morphology and functional variation for oil spill treatment. The urgent need for the advancements in day to day research made in this field and a broad range of applications of cellulose nanocrystals are also hereby discussed. The present and future challenges associated with the successful usage of these cellulosic nanocrystals for tackling this global phenomenon which is detrimental to marine ecosystem are also described.

## 2 Cellulose Nanocrystals and Their Preparation Procedures

Cellulose nanocrystals (CNCs) are rigid, highly crystalline nanoparticles having a morphology resembling a rod or a needle [6–8]. In addition to many benefits inherited from natural cellulose, such as low density, nontoxicity, biodegradability, renewability, and biocompatibility. A high aspect ratio, a rod-like morphology, chiral nematic activity, high mechanical capabilities, and flexible surface chemistry are some more unique characteristics of CNCs. Accordingly, the CNC parameters range from 100 to 1000 nm in length and 5–20 in diameter with aspect ratios of 10–120, depending on the cellulose supplies and hydrolysis techniques. Pure cellulose is transformed through a number both chemical and physical processes into CNCsA detailed explanation of the architecture, origin, and processing of cellulose is already well presented in the available literature of past several years [9–13]. The materials community has been particularly interested in cellulose nanocrystals (CNCs) because of their availability, inherent renewability, sustainability, and favorable physical and chemical properties [14] (Fig. 1).

The acid-to-cellulosic fiber ratio, in addition to the type and amount of acid used, all have an impact on CNC processing. Repeat centrifuge and dialysis are employed for several days for filter suspensions after acid treatment. Typically, sulfuric and hydrochloric acids are used to treat CNCs for acid hydrolysis. However, acid cleavage with phosphoric and hydrobromic acids has been found. When CNCs were hydrolyzed using hydrochloric acid, they flocculated and had a limited ability to scatter.

CNCs are created by acid hydrolysis of these cellulose fibers. Amorphous sections within the cellulose fibers were destroyed during in the acid hydrolysis, resulting in the formation of discrete crystalline domains known as CNCs. These plant-derived CNCs possess lengths ranging from 100 to 400 nm and lateral widths ranging from 5 to 70 nm. In order to create CNCs, pulp fibers can be hydrolyzed using a mineral acid, such as sulfuric, hydrochloric, and phosphoric acids. The type of acid utilized affects the precise functional group that is given on CNCs, which affects their crystalline stability. CNCs produced via sulfuric acid hydrolysis show extraordinary colloidal stability due to the electrostatic repulsion produced from negative charges sulfate ester groups on their surface [15].

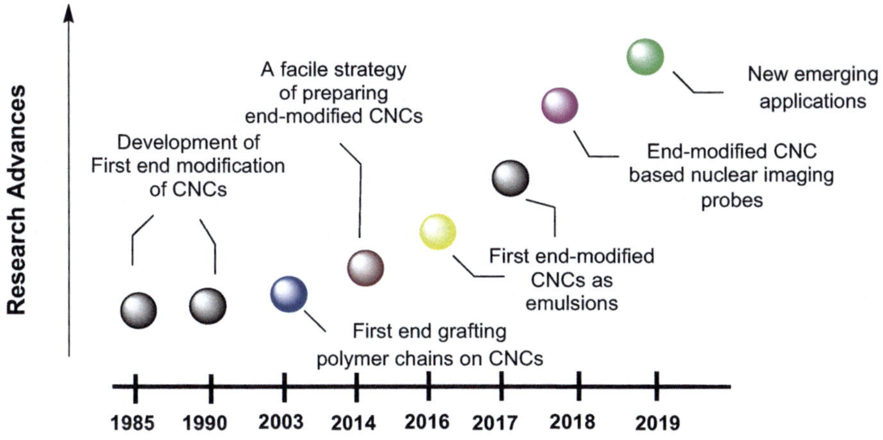


**Fig. 1** A timeline on the advancements in the studies of reducing end modifications of CNCs

Zhong et al. [16] investigated CNC colloidal stability inside the vicinity of organic and inorganic electrolyte ($Ca^{2+}$ and Na+ 140) (sodium dodecyl sulfate, sodium carboxymethyl cellulose, cationic polyacrylamide). They demonstrated that the negative charges sulphate esters groups on the surface of the CNCs repel each other electrostatically, causing the CNCs to be evenly spread in distilled water. However, the addition of appropriate organic and inorganic electrolytes may modify the stability of all these suspensions. Cherhal et al. 144 studied studied aggregation process for CNCs by raising or reducing overall surface charge density of CNCs. The adsorption capabilities of carboxylated CNCs generated in a single step of hydrolysis technique were investigated by He et al. [17] and Yu et al. [18]. Using ammonium persulfate to hydrolyze microcrystalline cellulose (MCC), He et al. created carboxylated CNCs (APS). The hydrolysis of MCC exposed the surface of CNCs with negatively charged carboxyl groups.

## 3 Morphology and Unique Properties of Cellulose Nanocrystals

Biodegradability, biocompatibility, as well as a high concentration of hydroxyl groups with adaptive structures are only a few of the beneficial characteristics of cellulose. It is classified as a carbohydrate polymer due to its chemical composition, which comprised of repeated –D46 glucopyranose units. Amorphous chains that are loosely packed and highly crystalline domains that are densely packed both make up cellulose fibers, and cellulose nanoparticles (CNs) with diverse morphologies and physical properties have been produced using various processing procedures [19]. Over the past ten years, a variety of nomenclatures have been employed to describe various types of cellulose-derived nanomaterials. These include bacterial cellulose, nanocrystalline cellulose, nanowhiskers, nanocrystals, nanocellulose, nanofibrils, nanofibers, and nanofibrillated cellulose, in addition to any combination of an above materials (Fig. 2).

TAPPI, the United States Department of Agriculture, the United States Forest Service, plus numerous Canadian institutions have standardized the classification of CNs into two major groups: cellulose nanocrystals (CNCs) and cellulose nanofibrils (CNFs) (CNFs). CNCs have several characteristics that make them a suitable choice to be used as flocculants used in water treatment. CNCs are nanoparticles with a large surface area that cause percolation in low concentrations. As a result, as compared to native fibers, CNCs can improve floc formation. CNCs' electrostatic repulsion and stiffness will prevent physical and chemical interactions, reducing gelation risk and boosting processability [20, 21].

According to Yu et al., CNCs made from MCC hydrolyzed in one step with citric acid and HCl performed better at flocculating dispersed kaolin clay particles over two other types of CNCs made by sulfuric acid & formic acid/HCl hydrolysis [22].

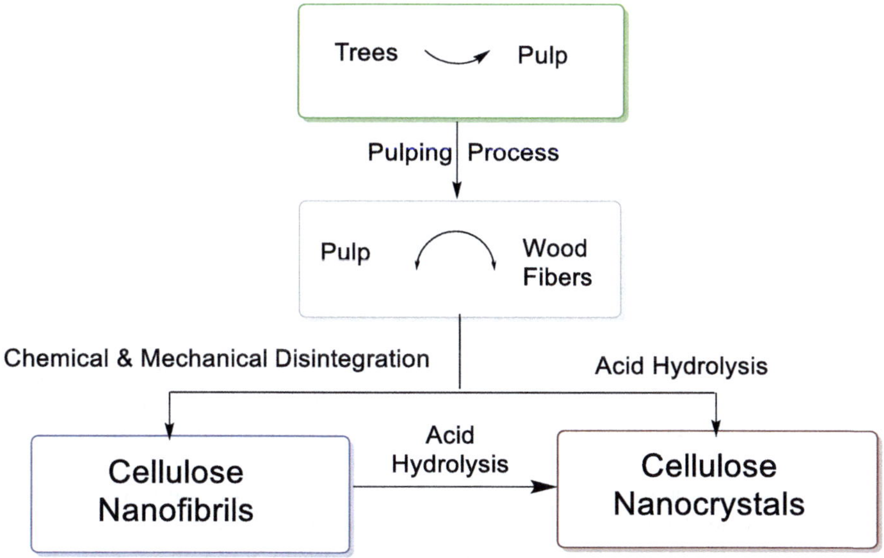


**Fig. 2** A basic schematic representation of process of extraction of cellulose nanocrystals from wood fibers

These discussed properties of cellulose nanocrystals are pre-requisite for its usage in various potential applications.

## 4 Utility of Cellulose Nanocrystals in Diverse Applications Due to Its Structural and Chemical Modifications

CNC is an abbreviation for cellulose nanocrystals CNC has emerged as a leading primary study concentrates on the cellulosic industry due to its biodegradability and effectiveness in increasing surface properties in composites, as well as its unique capabilities including liquid crystalline, optical, and self-assembly features [18, 23]. Meanwhile, cellulose nanocrystals has grown in popularity in a variety of fields due to its outstanding mechanical abilities and a lot of surface area, such as that used in packaging, drug distribution, cosmetics, photovoltaics, nanocomposites, and other things [24–27]. The discussion goes deeper into the current and potential uses of aligned CNC, including fillers to enhance the thermal and mechanical properties of biocomposites, photonic materials to enhance the optical features of iridescent films, including decreased water sorption for high resolution printing.

CNCs have such a wide range of applications in the realm of nanocomposites due to their intrinsic properties such as huge surface region and mechanical strength [28]. Because they have hydroxyl functional groups, they can be chemically functionalized and easily incorporated into a variety of polymer matrices [29–33]. These polymer composites are used in automotive parts, packaging, textiles, and coatings. To determine how cellulose nanocrystals interact with cells in vivo, a three-step chemical has been used to fluorescently label cellulose nanocrystals [34]. The very first study just on chemical grafting between aromatic molecules and polymer, known as the CNC reduction end group chemical modification, details complicated chemical events in the year 2003 [35].

## 5 Oil Spill Treatment and Its Present Strategies

Every year, inevitable oil spill incidents cause significant environmental damage to a ocean and coastal areas as well as the release of a significant amount of petroleum into the sea [36]. The health of the ecosystem's plants and animals would worsen if efficient oil spill response measures were not developed. The inclusion of dispersant is a workable approach of oil spill repair [37]. The Gulf of Mexico oil disaster required around 2.1 millions of gallons of chemical dispersant [38]. Surfactants and organic solvents are frequently combined to create common oil spill dispersants. These dispersants significantly reduce that surface tension among crude oil and saltwater, which allows spilt oil to break into small droplets when waves provide enough energy [39]. The process of an oil slick dispersing across liquid is essentially an emulsification.

A further benefit of using dispersants is that they produce tiny oil droplets, increasing the interfacial surface open to bacterial attack and promoting biodegradation [40]. But there is growing concern about their potential toxicity and environmental impact so because most of these surfactants are chemically produced [41].

Effective environmentally friendly products or technologies are necessary to protect the environment, especially aquatic environment, from of the detrimental effects of oil spills; else, ocean ecosystems were likely to suffer serious harm. Non-toxic, by products and various forms of biomasses have been generated at cheaper costs in order to be used for the introduction of micro- to nanoscale materials, membrane, aerogel and sponges, etc. can recover or extract oil from water resources [42].

Several events happen after an initial oil spill, including oil evaporation, oxidation, emulsification, sedimentation, biodegradation, and dispersion [43, 44]. Oil spills have been occurring for years, but sadly, cleanup techniques haven't changed all that much. Mechanical methods (such as booms, sorbents, and skimmers), thermal methods (such as in-situ burning), chemical methods (such as dispersion), and natural methods (such as navigation) are all used to clean up an oil spill (bioremediation) [45]. Each technique has advantages and disadvantages, but the efficiency of each commonly depends on external factors including oil composition, viscosity, the volume, as well as place and weather conditions [46].

Despite the enormous efforts made to regulate pollution, oil spills and the release of organic solvents can be disastrous because they pose a short-term threat to the lives of aquatic ecosystems sustain long-term, irreparable harm by thousands of species [47]. There are three sorts of cleanup methods of oil and organic solvents that are currently used. (i) Utilizing absorbents to remove oil and chemical solvent from of the water's surface; effectively dispersing oil and organic solvent to speed up natural deterioration; and also in burning off oil and organic solvent [48], The most effective and convenient method among these is the extraction of oil or an organic solvent using certain absorbents [49, 50].

Compressible aerogels are excellent materials for oil/organic solvent cleanup and recovery because of their extraordinary compaction, high porosity (up to 99.8%), low density (4–500 mg/cm$^3$), and high surface area (100–1000 m$^2$/g) [51, 52]. The mechanical robustness needs for compressive materials, however, are typically not met by conventional silica aerogels because they are frequently fragile and not sufficiently brittle [53]. Poly (vinyl alcohol) (PVA) aerogels have recently attracted interest because of their favourable characteristics, including affordability, biocompatibility, or biodegradability, in addition to their increased flexibility, decreased brittleness, and ease of manufacture [54–56]. However, PVA aerogels don't have a lot of mechanical strength or great compressibility. PVA aerogels supplemented with materials containing carbon nanotubes and graphene have recently been found to have good mechanical properties and great compressibility [57, 58]. These carbon-based compounds, however, are difficult to obtain for use in large-scale applications because of the high cost of raw materials.

## 6 Applications and Advancements in the Use and Study of Cellulose Nanocrystals for Oil Spill Treatment

Since over three decades, there have been several oil spill events, which are caused by various oil transporters. One of the main contributors to environmental pollution are these unintentionally spilled oils. We cannot prevent accidental oil leaks from occurring as a result of mechanical failure, transportation, utility failure, and improper storage conditions [2].

Major occupational, environmental, and public health catastrophes are frequently caused by oil spills. If proper spill remediation processes are not used, it is possible for crude oil spills to harm the ecosystem's plants and animals. Surface-active agents or dispersants can be used to break up large oil spills into smaller ones, which is a practical way to handle oil spills. Surface-active compounds that can hasten that biodegradation at the oil–water interface include dispersants and lipophilic fertilizers [3–5, 59].

Ojala et al. [60] investigated the composition and properties of oil-in-water (O/W) marine diesel emulsions stabilized by bifunctionalized cellulose nanocrystals. The study discovered that bifunctionalized CNCs might lower the size of oil droplets in emulsions while also acting as stabilizers.

Pandey et al. [61] described an oil-in-water emulsion stabilized using cellulose nanocrystals (CNCs) of two different surface charge levels and aggregation states. Their findings revealed that the a-CNC-stabilized emulsion had faster adsorption kinetics due to a lower electrostatic barrier between the particles and the oil/water interface, smaller droplet size due to the higher aspect ratio of aggregated a-CNCs, lower interface coverage, multiple layers of CNCs at the interface, and a small amount of unadsorbed CNCs in the continuous phase.

CNCs can be used to create nanopapers and nanocoatings with high anti-crude oil properties. These materials can successfully separate crude oil from water, even under difficult conditions such as high viscosity and adhesion. The nanopapers generated using vacuum filtering can separate surfactant-stabilized crude oil-in-water emulsions with an amazing water flow of up to 35,000 L m without substantial degradation in performance over time [62].

CNCs have been investigated as surface-washing agents to clean polluted shorelines. Their low cost and non-toxic characteristics make them ideal for environmental applications that prioritize safety and efficiency. This use is especially useful for dealing with oil spills on beaches and in coastal locations, where standard approaches may be less effective [63].

Hydrophobic nanocellulose aerogels have been developed recently; however, they need to be modified in order to improve their efficiency in oil/water separation. These aerogels are lightweight and eco-friendly, yet they have a large oil absorption capacity. Their surface qualities can be changed, enabling customization according to certain spill conditions [64].

Composite nanofibers combining cellulose acetate and cellulose nitrate with carbon black have been developed to improve oil removal effectiveness. These composites show excellent absorption capacity, reusability, and stability [65]. The nanofiber mats can absorb more than their weight in oil in 30 min, showing rapid response capabilities that are critical during spills.

The biodegradability of CNCs, as opposed to conventional synthetic adsorbents, reduces environmental effect and is one of the main benefits of employing them. This characteristic is crucial for environmentally friendly methods of managing oil spills [66, 67].

## 7 Prime Difficulties in the Implementation of Remedial Measures for Oil Spill Treatment

Bioremediation is a process where oil spills are used as the main energy source by marine microorganisms. Recent research suggests that adding nanoparticles to oil-contaminated areas could increase the efficiency with which oil is broken by marine microorganisms. The majority of the time, nanoparticles are utilized for magnetic sorbents, but they may also function as emulsifiers, increasing overall bioavailability of a oil by giving bacteria a surface (droplets) to adhere to and encourage growth.

Oil spills cause water bodies to become contaminated, which is what pollution is. To improve the state of the planet oil leaks from vessels, tankers, and marine oil pipes must be fixed. To protect the ecology and water, this oil must be taken out of the water or surface. We have observed a significant shift in the pattern with the drop in the number of tanker and pipes for such transport of oil and petroleum derivatives, which has lessened the chance of oil spills. The crucial point would be that, despite the reality that oil spills obviously garner greater media attention, researchers assert that other types of pollution hurt the marine ecosystem significantly more. Oil spill cleanup requires the use of contemporary, ecologically safe tools immediately.

According to certain studies, using organic material in the event of an oil leak could be a successful strategy. Oil leaks and spills harm the environment for a long time, especially the neighboring marine ecosystems. Even though it is crucial to remove contaminated oil quickly and safely, remediation techniques have not improved much over the past 20 years or more. It might be challenging to use removal techniques like skimmers, on situ burning, and sorbents in bad weather. Bioremediation is a process where oil spills are used as the main energy source by marine microorganisms. Recent research suggests that adding nanoparticles to oil-contaminated areas could increase the efficiency with which oil marine microorganisms degrade.

Nanoparticles have primarily been utilized in magnetic sorbents, but they can also function as emulsifiers, increasing oil bioavailability through giving bacteria with such a surface (droplets) to adhere to and stimulate growth. Emulsification increases its oil–water interfacial area, resulting in more oil being broke down by microorganisms. On a range of oil compositions, nanoenhanced bioremediation works well and is barely affected by the weather. It provides a safe substitute for conventional remediation techniques, such as inefficient and frequently harmful chemical dispersants, sorbents, and in situ burning.

# 8 Conclusions

The features of cellulose nanocrystals (CNCs) have been the subject of numerous studies with the aim of regulating them for a number of potential uses. The most of CNC modification procedures, however, focus on surface modification because of the availability and accessibility of numerous hydroxyl groups that are present on the surface. A number of studies have indeed been published in this field, and only recently has the topic of reducing CNC finishing modification started to attract more attention. One of the essential elements of end modification approaches is the specific-site change just at terminal area while maintaining the surface properties or properties of CNCs.

An innovative technique for the cleanup of oil spills is indeed the releasing of surfactants via interfacially active hollow particles. The release of surfactant molecules reduces the tension at the oil–water interface, and the adsorption of halloysite nanotubes (HNT) creates a steric barrier to drop coalescence. The nanotubes are also discovered to assume a side-on configuration and create end-to-end links to build networks on the interface. Oil droplets produced by the cooperative stabilisation of emulsification via halloysite and surfactants are smaller than those produced by halloysite acting as a particulate emulsifier alone. Measurements of the interfacial tension between the crude oil and the water reveal that the use of the surfactant-loaded-HNT dispersant reduces it to values suitable for the dispersion of oil spills. When used in practical settings, the dispersion of spilled oil into microscopic droplets will reveal a sizable oil-seawater interfacial surface, allowing for the final bioremediation for oil spills by marine native microorganisms.

Because these materials remain stable at the oil–water interface, we believe that halloysite stabilized emulsions have had the potential for a variety of further uses in oil spill cleaning. It is also feasible to create nanotubes that do not only contain surfactants but also hydrophobic fluorescent indicators which will break into the oil phase and help with oil spill monitoring at night. The nanotubes can be enhanced with magnetic indications to aid mobilize oil spills. These are relevant to several areas in our ongoing research. The capacity to obtain halloysite nanotubes in vast quantities from natural sources also opens the door to the possibility of large-scale industrial uses.

## References


1. Ivshina IB, Kuyukina MS, Krivoruchko AV, Elkin AA, Makarov SO, Cunningham CJ, Peshkur TA, Atlas RM, Philp JC (2015) Oil spill problems and sustainable response strategies through new technologies. Environ Sci Process Impacts 17:1201
2. Goldstein BD, Osofsky HJ, Lichtveld MY (2011) Current concepts the gulf oil spill. New Engl J Med 364(14):1334–1348
3. Atlas RM (1995) Petroleum biodegradation and oil spill bioremediation. Mar Pollut Bull 31(4–12):178–182
4. Venkataraman P, Tang JJ, Frenkel E, McPherson GL, He JB, Raghavan SR, Kolesnichenko V, Bose A, John VT (2013) Attachment of a hydrophobically modified biopolymer at the oil-water interface in the treatment of oil spills. Acs Appl Mater Inter 5(9):3572–3580
5. Lessard RR, Demarco G (2000) The significance of oil spill dispersants. Spill Sci Technol B 6(1):59–68
6. Dufresne A (2019) Curr For Rep 5:76–89. https://doi.org/10.1007/s40725-019-00088-1
7. Dufresne A (2017) Nanocellulose: from nature to high performance tailored materials, 2nd ed. Walter de Gruyter GmbH: Berlin/Boston. https://doi.org/10.1515/9783110480412-001
8. Thomas B, Raj MC, Joy J, Moores A, Drisko GL, Sanchez C (2018). Chem Rev 118:11575–11625. https://doi.org/10.1021/acs.chemrev.7b00627
9. Nimz HH (1984) Wood-chemistry, ultrastructure, reactions. Holz als Roh- und Werkst 42(8):314. https://doi.org/10.1007/BF02608943
10. Yang H, Yan R, Chen H, Lee DH, Zheng C (2007) Characteristics of hemicellulose, cellulose and lignin pyrolysis. Fuel 86(12–13):1781–1788. https://doi.org/10.1016/J.FUEL.2006.12.013
11. Marchessault RH, Malhotra SL, Jones AY, Perovic A (1983) The wood explosion process: characterization and uses of lignin/cellulose products. In: Soltes EJ (ed) Wood a agricultural residues. Academic Press, pp 401–413. https://doi.org/10.1016/B978-0-12-654560-9.50026-3
12. Chang T-C, Chang H-T, Wu C-L, Chang S-T (2010) Influences of extractives on the photodegradation of wood. Polym Degrad Stab 95(4):516–521. https://doi.org/10.1016/J.POLYMDEGRADSTAB.2009.12.024

13. Gralen N, Svedberg THE (1943) Molecular weight of native cellulose. Nature 152(3865):625. https://doi.org/10.1038/152625a0
14. Habibi Y, Lucia LA, Rojas OJ (2010) Cellulose nanocrystals: chemistry, self-assembly, and applications. Chem Rev 110(6):3479–3500. https://doi.org/10.1021/cr900339w
15. Klemm D, Kramer F, Moritz S, Lindstrom T, Ankerfors M, Gray D, Dorris A (2011). Angew Chemie Int Ed50:5438–5466
16. Zhong L, Fu S, Peng X, Zhan H, Sun R (2012) Carbohydr Polym 90:644–649
17. He X, Male KB, Nesterenko PN, Brabazon D, Paull B, Luong JHT (2013) ACS ApplMater Interfaces 5:8796–8804
18. Yu H-Y, Zhang D-Z, Lu F-F, Yao J, Sustain ACS (2016) Chem Eng 4(2632–1562):2643
19. Peng BL, Dhar N, Liu HL, Tam KC (2011) Can J Chem Eng 89:1191–1206
20. Carpenter AW, De Lannoy CF, Wiesner MR (2015). Environ Sci Technol 49(1511):5277–5287
21. Vandamme D, Eyley S, Van den Mooter G, Muylaert K, Thielemans W (2015) Bioresour Technol 194:270–275
22. Eyley S, Vandamme D, Lama S, Van den Mooter G, Muylaert K, Thielemans W (2015) Nanoscale 7:14413–14421
23. Dufresne A (2012) Nanocellulose: from nature to high performance tailored materials. De Gruyter. Nanocellulose. https://doi.org/10.1515/9783110254600
24. Thomas B, Raj MC, Joy J, Moores A, Drisko GL, Sanchez C (2018) Nanocellulose, a versatile green platform: from biosources to materials and their applications. Chem Rev 118(24):11575–11625. https://doi.org/10.1021/acs.chemrev.7b00627
25. De France KJ, Hoare T, Cranston ED (2017) Review of hydrogels and aerogels containing nanocellulose. Chem Mater 29(11):4609–4631. https://doi.org/10.1021/acs.chemmater.7b00531
26. Kim J-H, Lee D, Lee Y-H, Chen W, Lee S-Y (2019) Nanocellulose for energy storage systems: beyond the limits of synthetic materials. Adv Mater 31(20):1804826. https://doi.org/10.1002/adma.201804826
27. Chen W, Yu H, Lee S-Y, Wei T, Li J, Fan Z (2018) Nanocellulose: a promising nanomaterial for advanced electrochemical energy storage. Chem Soc Rev 47(8):2837–2872. https://doi.org/10.1039/C7CS00790F
28. Abdul Rashid ES, Muhd Julkapli N, Yehye WA (2018) Nanocellulose reinforced as green agent in polymer matrix composites applications. Polym Adv Technol 29(6):1531–1546. https://doi.org/10.1002/pat.4264
29. Favier V, Chanzy H, Cavaille JY (1995) Polymer nanocomposites reinforced by cellulose whiskers. Macromolecules 28(18):6365–6367. https://doi.org/10.1021/ma00122a053
30. Samir MASA, Alloin F, Sanchez J-Y, Dufresne A (2004) Cellulose nanocrystals reinforced poly(oxyethylene). Polymer (Guildf) 45(12):4149–4157. https://doi.org/10.1016/j.polymer.2004.03.094
31. Samir MASA, Chazeau L, Alloin F, Cavaillé J-Y, Dufresne A, Sanchez J-Y (2005) POE-based nanocomposite polymer electrolytes reinforced with cellulose whiskers. Electrochim Acta 50(19):3897–3903. https://doi.org/10.1016/j.electacta.2005.02.065
32. de Menezes AJ, Siqueira G, Curvelo AAS, Dufresne A (2009) Extrusion and characterization of functionalized cellulose whiskers reinforced polyethylene nanocomposites. Polymer (Guildf) 50(19):4552–4563. https://doi.org/10.1016/j.polymer.2009.07.038
33. Qi H, Cai J, Zhang L, Kuga S (2009) Properties of films composed of cellulose nanowhiskers and a cellulose matrix regenerated from alkali/urea solution. Biomacromol 10(6):1597–1602. https://doi.org/10.1021/bm9001975
34. Petersson L, Mathew AP, Oksman K (2009) Dispersion and properties of cellulose nanowhiskers and layered silicates in cellulose acetate butyrate nanocomposites. J Appl Polym Sci 112(4):2001–2009. https://doi.org/10.1002/app.29661
35. Dong S, Roman M (2007) Fluorescently labeled cellulose nanocrystals for bioimaging applications. J Am Chem Soc 129(45):13810–13811. https://doi.org/10.1021/ja0761961
36. Sipahi-Sağlam MG, Gruber E (2013). Cellulose 10:237–250. https://doi.org/10.1023/A:1025151701985

37. Etkin DS (2001) Analysis of oil spill trends in the United States and worldwide. In: Proceedings of the 2001 international oil spill conference, Tampa, FL. American Petroleum Institute, Washington, DC, pp 1291–1300
38. Chapman H, Purnell K, Law RJ, Kirby MF (2007) The use of chemical dispersants to combat oil spills at sea: a review of practice and research needs in Europe. Mar Pollut Bull 54(7):827–838
39. Allan SE, Smith BW, Anderson KA (2012) Impact of the deepwater horizon oil spill on bioavailable polycyclic aromatic hydrocarbons in gulf of mexico coastal waters. Environ Sci Technol 46(4):2033–2039
40. Lessard RR, DeMarco G (2000) The significance of oil spill dispersants. Spill Sci Technol Bull 6(1):59–68
41. Churchill PF, Dudley RJ, Churchill SA (1995) Surfactant enhanced bioremediation. Waste Manage 15(5–6):371–377
42. Almeda R, Hyatt C, Buskey EJ (2014) Toxicity of dispersant Corexit 9500A and crude oil to marine microzooplankton. Ecotoxicol Environ Saf 106:76–85
43. Fingas M (2012) The basics of oil spill cleanup, 3rd edn. Lewis Publ, Boca Raton
44. Passow U (2016) Formation of rapidly-sinking, oil-associated marine snow. Deep Sea Res. Part II 129:232–240
45. United States Environmental Protection Agency (2016) The fate of spilled oil. https://archive.epa.gov/emergencies/content/learning/web/html/oilfate.html. Accessed 3 Jan 2020
46. Dave D, Ghaly AE (2011) Remediation technologies for marine oil spills: a critical review and comparative analysis. Am J Environ Sci 7:424–440
47. United States Environmental Protection Agency (1999) Understanding oil spills and oil spill response. https://www.epa.gov/emergencyresponse/understanding-oil-spills-and-oil-spill-response. Accessed 4 April 2020
48. Sai HZ, Fu R, Xing L, Xiang JH, Li ZY, Li F, Zhang T (2015) Surface modification of bacterial cellulose aerogels' web-like skeleton for oil/water separation. Acs Appl Mater Inter 7(13):7373–7381. https://doi.org/10.1021/acsami.5b00846
49. Yang Y, Deng YH, Tong Z, Wang CY (2014) Multifunctional foams derived from poly(melamine formaldehyde) as recyclable oil absorbents. J Mater Chem A 2(26):9994–9999. https://doi.org/10.1039/c4ta00939h
50. Duan B, Gao HM, He M, Zhang LN (2014) Hydrophobic modification on surface of chitin sponges for highly effective separation of oil. Acs Appl Mater Inter 6(22):19933–19942. https://doi.org/10.1021/am505414y
51. Heath L, Thielemans W (2010) Cellulose nanowhisker aerogels. Green Chem 12(8):1448–1453. https://doi.org/10.1039/c0gc00035c
52. Wu ZY, Li C, Liang HW, Chen JF, Yu SH (2013) Ultralight, flexible, and fire-resistant carbon nanofiber aerogels from bacterial cellulose. Angew Chem Int Edit 52(10):2925–2929. https://doi.org/10.1002/anie.201209676
53. Yang X, Cranston ED (2014) Chemically cross-linked cellulose nanocrystal aerogels with shape recovery and superabsorbent properties. Chem Mater 26(20):6016–6025. https://doi.org/10.1021/cm502873c
54. Stammen JA, Williams S, Ku DN, Guldberg RE (2001) Mechanical properties of a novel PVA hydrogel in shear and unconfined compression. Biomaterials 22(8):799–806. https://doi.org/10.1016/S0142-9612(00)00242-8
55. Hou Y, Chen C, Liu KM, Tu Y, Zhang L, Li YB (2015) Preparation of PVA hydrogel with high-transparence and investigations of its transparent mechanism. Rsc Adv 5(31):24023–24030. https://doi.org/10.1039/c5ra01280e
56. Hassan CM, Peppas NA (2000) Structure and morphology of freeze/thawed PVA hydrogels. Macromolecules 33(7):2472–2479. https://doi.org/10.1021/Ma9907587
57. Sui ZY, Meng QH, Zhang XT, Ma R, Cao B (2012) Green synthesis of carbon nanotube-graphene hybrid aerogels and their use as versatile agents for water purification. J Mater Chem 22(18):8767–8771. https://doi.org/10.1039/c2jm00055e
58. Kabiri S, Tran DNH, Altalhi T, Losic D (2014) Outstanding adsorption performance of graphene-carbon nanotube aerogels for continuous oil removal. Carbon 80:523–533. https://doi.org/10.1016/j.carbon.2014.08.092

59. Choi HM, Cloud RM (1992) Natural sorbents in oil spill cleanup. Environ Sci Technol 26(4):772–776
60. Lvov YM, Shchukin DG, Mohwald H, Price RR (2008) Halloysite clay nanotubes for controlled release of protective agents. ACS Nano 2(5):814–820
61. Ojala J, Sirviö JA, Liimatainen H (2016) Nanoparticle emulsifiers based on bifunctionalized cellulose nanocrystals as marine diesel oil–water emulsion stabilizers. Chem Eng J 288:312–320
62. Pandey A, Derakhshandeh M, Kedzior SA, Pilapil B, Shomrat N, Segal-Peretz T, Bryant SL, Trifkovic M (2018) Role of interparticle interactions on microstructural and rheological properties of cellulose nanocrystal stabilized emulsions. J Colloid Interface Sci 532:808–818
63. Wu M, Zhang C, Pi J, Liu C, Yang J, Xu Z (2019) Cellulose nanocrystals as anti-oil nanomaterials for separating crude oil from aqueous emulsions and mixtures. J Mater Chem A
64. Rengyu Yue Chunjiang An Zhibin Ye Shan Zhao (2021) Exploring the use of cellulose nanocrystal as surface-washing agent for oiled shoreline cleanup. J Hazard Mater 402:123464
65. Chhajed M, Verma C, Maji PK (2024) Recent advances in hydrophobic nanocellulose aerogels for oil spill applications: a review. Mar Pollut Bull 199:116024. https://doi.org/10.1016/j.marpolbul.2024.116024
66. Elmaghraby NA, Omer AM, Kenawy ER et al (2022) Fabrication of cellulose acetate/cellulose nitrate/carbon black nanofiber composite for oil spill treatment. Biomass Conv Bioref. https://doi.org/10.1007/s13399-022-03506-w
67. Fürtauer S, Hassan M, Elsherbiny A, Gabal SA, Mehanny S, Abushammala H (2021) Current status of cellulosic and nanocellulosic materials for oil spill cleanup. Polymers 13:2739. https://doi.org/10.3390/polym13162739
68. Trinh TN, Loc ND, Nam TV (2023) Modified methods of oil cleanup with cellulose–based adsorbents: a review. VN J Hydrometeorol 14:96–120

GPSR Compliance

*The European Union's (EU) General Product Safety Regulation (GPSR) is a set of rules that requires consumer products to be safe and our obligations to ensure this.*

*If you have any concerns about our products, you can contact us on ProductSafety@springernature.com*

In case Publisher is established outside the EU, the EU authorized representative is:

Springer Nature Customer Service Center GmbH
Europaplatz 3
69115 Heidelberg, Germany

**Batch number: 10725825**

Printed by Printforce, the Netherlands